Synthesis Lectures on Mechanical Engineering

This series publishes short books in mechanical engineering (ME), the engineering branch that combines engineering, physics and mathematics principles with materials science to design, analyze, manufacture, and maintain mechanical systems. It involves the production and usage of heat and mechanical power for the design, production and operation of machines and tools. This series publishes within all areas of ME and follows the ASME technical division categories.

Edward B. Magrab

Static Analysis of Beams, Frames, and Arches

A Direct Approach

Springer

Edward B. Magrab
Department of Mechanical Engineering
(emeritus), University of Maryland
College Park, MD, USA

ISSN 2573-3168 ISSN 2573-3176 (electronic)
Synthesis Lectures on Mechanical Engineering
ISBN 978-3-032-28725-0 ISBN 978-3-032-28726-7 (eBook)
https://doi.org/10.1007/978-3-032-28726-7

This Springer imprint is published by the registered company Springer Nature Switzerland AG
The registered company address is: Gewerbestrasse 11, 6330 Cham, Switzerland

In memory of June Coleman Magrab

Preface

The objective of this work is to present a direct and unified approach to determine the response of beam structures to static loads when the beams can be modeled using the Euler-Bernoulli theory. The approach takes the Laplace transform of the governing ordinary differential equation with respect to the spatial variable. The Laplace transform was selected because it easily handles the delta function and the unit step function. In addition, in many instances, one is able to immediately set to zero half the number of solution constants that must be determined from the boundary conditions, thereby making the subsequent algebraic manipulations practical.

In Chap. 1, a synopsis of the material in Chaps. 2–6 is given to show how the solution for the constant cross-section beam is used to obtain solutions for beams with a step change in flexural rigidity, rectangular frames, and rectangular plates hinged on two opposite edges. In Chap. 2, a solution is obtained for a uniform beam subject to several factors and several applied loads acting simultaneously for different boundary conditions. The factors consist of the presence or absence of an axial force, an elastic foundation, and rigid in-span supports. The applied loads are composed of the presence or absence of distributed forces, concentrated forces, and moments. The boundary conditions are either clamped-clamped, clamped-hinged, clamped-free, or hinged-hinged. These solutions are then used directly in Chap. 3 to obtain solutions to beams with a step change in flexural rigidity and, in Chap. 4, to rectangular frames with three, five, and six beams when these systems are subjected simultaneously to many of the same factors and applied loads as the uniform beam. It will be seen that this approach provides closed form solutions for several new results: uniform and stepped beams with rigid in-span supports, essentially multi-span beams, and the spatial distribution of the frame's displacement. In addition, for the uniform beam and the beam with a step change in flexural rigidity, the lowest buckling loads are determined with or without in-span rigid supports. It is then shown in Chap. 5 that for a rectangular plate hinged on two opposite edges, the governing equation reduces to that of a uniform beam; consequently, the solution for the uniform beam can be used directly. Lastly, in Chap. 6, it is shown how this approach can be used to obtain the response of asymmetric uniform circular arches subjected to various asymmetric loads.

For most of the analytical results presented here, numerical values can be obtained from a set of five interactive graphics programs that can be downloaded from the SpringerLink page for each chapter. The link's address also appears on the first page of each of Chaps. 2–6. These programs provide values in real-time making parametric and "what-if" investigations practical. They can be used to explore new results that are unavailable elsewhere, verify solutions to exercises and examples found in engineering texts, and to evaluate and extend textbook/handbook material to explore more complex models and loading combinations. All figures giving numerical examples for the various systems were generated with these programs.

College Park, MD, USA Edward B. Magrab

Competing Interests The author has no competing interests to declare that are relevant to the content of this manuscript.

Contents

Introduction

1

Abstract

The three sets of factors that influence the form of the solution to the response of beams to static forces are discussed. This is followed by a synopsis of the governing equations and how the solution for the constant cross-section beam is used to obtain the solutions to the subsequent elastic systems.

Keywords

Beams · Axial force · Elastic foundation · Rigid supports · Concentrated force · Distributed force · Moments · Boundary conditions

1.1 Introduction

Beams are ubiquitous in engineering applications, and their importance cannot be overstated. The analysis of the response of different beam configurations to various types of static loads using the Euler-Bernoulli beam theory has been analyzed in many books using a wide variety of techniques. For examples, see textbooks [1, 2] and reference works [3–8]. Although detailed analysis requires more sophisticated solution techniques, such as the finite element method, preliminary analyses using closed-form solutions from the Euler-Bernoulli theory are still very useful for exploring initial configurations and the effects of parameter values.

Supplementary Information The online version contains supplementary material available at (https://doi.org/10.1007/978-3-032-28726-7_1).

E. B. Magrab, *Static Analysis of Beams, Frames, and Arches*, Synthesis Lectures on Mechanical Engineering, https://doi.org/10.1007/978-3-032-28726-7_1

A beam system is influenced by three sets of factors that are often the reasons why different solution methods are used. The first set of factors consists of the presence or absence of an axial force, an elastic foundation, and rigid in-span supports. The beam's boundary conditions comprise the second set of factors: typically, clamped-clamped, clamped-hinged, clamped-free, and hinged-hinged. The third set of factors are the applied loads composed of the presence or absence of distributed forces, concentrated forces, and moments.

Rather than considering a variety of methods to solve various beam configurations, we shall show that the approach presented in the subsequent chapters and summarized in Sect. 1.2 is sufficient to determine the spatial distribution of the displacement, rotation, moment, and shear force response of the following systems:

1. Constant cross-section beams with in-span rigid supports [Chap. 2]
2. Beams with a step change in flexural rigidity with in-span rigid supports [Chap. 3]
3. Three-beam rectangular frames with their grounded ends having any combination of fixed (clamped) and hinged and for five-beam and six-beam rectangular frames with fixed ends [Chap. 4]
4. Rectangular plate with two opposite edges hinged; illustration of the reduction of the governing equation of a two-dimensional elastic system into a one-dimensional equation that is identical to that of a uniform beam [Chap. 5]
5. Asymmetrical circular arches with either fixed or hinged ends and subject to various forces; illustration of the approach to a system with two coupled governing differential equations [Chap. 6]

The factors that are considered simultaneously in the solutions are summarized below. In most cases, for any combination of these factors, the displacement, rotation, moment, and shear force as a function of position are determined. By generating these general solutions, one can easily examine their special cases.

System 1—Uniform Beam

Factors: elastic foundation, compressive axial force, and none, one, or two in-span rigid supports

Boundary conditions: clamped-clamped, clamped-hinged, clamped-free, or hinged-hinged

Applied loads: simultaneous application of one, two, or three nonoverlapping distributed forces, one, two, or three concentrated forces, and one, two, or three moments

Buckling load: elastic foundation and none, one, or two in-span rigid supports

System 2—Beam with Step Change in Flexural Rigidity

Factors: elastic foundation, compressive axial force, and none or one in-span rigid support in each beam segment

Boundary conditions: clamped-clamped, clamped-hinged, clamped-free, or hinged-hinged

Applied loads: for each beam segment the simultaneous application of one distributed force, one concentrated force, and one moment

Buckling loads: elastic foundation and none or one in-span rigid support in each beam segment

System 3—Frames

Boundary conditions: For the grounded ends, any combination of clamped and hinged.

Applied loads: On each frame's beam, there can be one distributed force and one concentrated force.

Attributes: Any reasonable combination of length and flexural rigidity of each beam.

System 4—Plates Hinged on Two Opposite Edges

Factors: elastic foundation and independent compressive axial forces parallel to the respective plate edges

Boundary conditions: two opposite edges hinged, the third edge either clamped or hinged, and the fourth edge either clamped, hinged, or free

Applied loads: distributed load constant in the direction parallel to the opposing hinged edges and linearly varying in the other direction

System 5—Circular Arch

Boundary conditions: hinged or clamped at each end

Applied loads: nonoverlapping distributed constant vertical loads on portions of the arch, gravity load, and nonoverlapping distributed constant radial loads on portions of the arch

1.2 Synopsis of the Solution Method

We shall now give a preview of the method that provides a direct approach to obtaining the respective responses of the five systems mentioned in Sect. 1.1 to applied loads. For the definitions of the terms appearing in the subsequent equations and other details, see the applicable chapter.

The method employs the straightforward application of the Laplace transform to the spatial variable of the system's governing equation(s). This method has been chosen for its ability to solve ordinary differential equations with constant coefficients that contain step functions and delta functions. In addition, the boundary conditions that are considered will, in most cases, immediately halve the number of solution constants. This will lead to a simplification of the subsequent manipulations, making it possible to get algebraic results not previously available, for example, uniform and stepped beams with in-span rigid supports. It will also be seen that by keeping the solution in functional form, one can get a general solution that includes all factors and all types of applied forces appearing simultaneously. In other words, we treat these five systems as five different boundary value problems using one general purpose solution method (the Laplace transform) and one general solution, that of a uniform beam. This has the advantage that one does not have to make any simplifying assumptions regarding the number and types of loads that can be considered and eliminates the need to make assumptions about its shape, boundary conditions, or loading.

System 1 The governing equation in terms of nondimensional quantities is

$$\begin{aligned}\frac{d^4w}{d\eta^4}+S_o\frac{d^2w}{d\eta^2}+K_fw+\sum_{j=1}^{2}K_j\delta(\eta-\eta_j)w=\sum_{j=1}^{3}g_j(\eta)\\+\sum_{j=1}^{3}F_{c,j}\delta(\eta-\eta_{c,j})-\sum_{j=1}^{3}T_j\frac{d}{d\eta}\delta(\eta-\eta_{T,j})\end{aligned}\tag{1.1}$$

where $\delta(\eta)$ is the delta function, S_o represents the compressive force, K_f is the elastic foundation, K_j is the in-span elastic springs, $F_{c,j}$ is the magnitude of the concentrated forces, T_j is the magnitude of the moments, and $g_j(\eta)$ describes the distributed forces. The elastic springs are introduced to provide the means of becoming rigid in-span supports. At the appropriate point in the solution process, their magnitudes are set to infinity. This results in an analytical expression for the displacement of a beam with two arbitrarily positioned in-span rigid supports.

The solution to Eq. (1.1) is obtained using the Laplace transform, which yields

$$\begin{aligned}w(\eta)=d'(\eta)w(0)+d(\eta)w'(0)+v'(\eta,0)w''(0)\\+v(\eta,0)w'''(0)+H(\eta)-K(\eta)\end{aligned}\tag{1.2}$$

where

$$\begin{aligned}H(\eta)&=\sum_{j=1}^{3}h_j(\eta)+\sum_{j=1}^{3}F_{c,j}\Phi(\eta,\eta_{c,j})-\sum_{j=1}^{3}T_j\Psi(\eta,\eta_{T,j})\\K(\eta)&=\sum_{j=1}^{2}K_jw(\eta_j)\Phi(\eta,\eta_j)\end{aligned}\tag{1.3}$$

and d, v, Φ, and Ψ are solution functions, the prime denotes the derivative with respect to η, and h_j is the solution from the application of the distributed forces. In Eq. (1.2), $w(0)$ is the displacement at $\eta = 0$, $w'(0)$ is the rotation of the cross section at $\eta = 0$, $w''(0)$ is proportional to the bending moment at $\eta = 0$, and $w'''(0)$ is proportional to the shear force at $\eta = 0$. These are the solution constants that are determined from the boundary conditions. Assuming that the end of the beam at $\eta = 0$ is either clamped or hinged, two of these solution constants vanish, and the remaining two are obtained from satisfying the boundary conditions at $\eta = 1$, which are either clamped, free, or hinged. This immediate reduction in the number of solution constants is what makes the subsequent algebra presented in Chaps. 2–6 feasible.

The inclusion of the in-span rigid supports in the general solution obtained in the manner shown appears to be a new result.

System 2 The governing equations for each segment of the beam in terms of nondimensional quantities are

$$\begin{aligned}
&\frac{d^4 w_1}{d\eta^4} + S_o \frac{d^2 w_1}{d\eta^2} + K_f w_1 + K_1 \delta(\eta - \eta_1) w_1 = g_1(\eta) \\
&\qquad + F_{c,1}\delta(\eta - \eta_{c,1}) - T_1 \frac{d}{d\eta}\delta(\eta - \eta_{T,1}) \quad 0 \le \eta \le a_s \\
&\frac{d^4 w_2}{d\xi^4} + \beta S_o \frac{d^2 w_2}{d\xi^2} + \beta K_f w_2 + \beta K_2 \delta(\xi - \xi_2) w_2 = \beta g_2(\xi) \\
&\qquad + \beta F_{c,2}\delta(\xi - \xi_{c,2}) - \beta T_2 \frac{d}{d\xi}\delta(\xi - \xi_{T,2}) \quad 0 \le \xi \le b_s
\end{aligned} \tag{1.4}$$

where β is the ratio of the flexural rigidity of the segments, a_s is the nondimensional length of segment 1, and $b_s = 1 - a_s$ is the nondimensional length of segment 2. In Eq. (1.4), the origin of segment 1 is at the left end of the beam, and the origin of segment 2 is at the right end of the beam. We see from Eqs. (1.1) to (1.3) that the solutions to Eq. (1.4) are

$$\begin{aligned}
w_1(\eta) &= d'(\eta) w_1(0) + d(\eta) w_1'(0) + v'(\eta, 0) w_1''(0) \\
&\quad + v(\eta, 0) w_1'''(0) + H_1(\eta) - K_{s,1}(\eta) \quad 0 \le \eta \le a_s \\
w_2(\xi) &= \hat{d}'(\xi) w_2(0) + \hat{d}(\xi) w_2'(0) + \hat{v}'(\xi, 0) w_2''(0) \\
&\quad + \hat{v}(\xi, 0) w_2'''(0) + \beta \hat{H}_2(\xi) - \beta \hat{K}_{s,2}(\xi) \quad 0 \le \xi \le b_s
\end{aligned} \tag{1.5}$$

where

$$\begin{aligned}
H_1(\eta) &= h_1(\eta) + F_{c,1}\Phi(\eta, \eta_{c,1}) - T_1 \Psi(\eta, \eta_{T,1}) \\
\hat{H}_2(\xi) &= \hat{h}_2(\xi) + F_{c,2}\hat{\Phi}(\xi, \xi_{c,2}) - T_2 \hat{\Psi}(\xi, \xi_{T,2}) \\
K_{s,1}(\eta) &= K_1 w_1(\eta_1) \Phi(\eta, \eta_1) \\
\hat{K}_{s,2}(\xi) &= K_2 w_2(\xi_2) \hat{\Phi}(\xi, \xi_2)
\end{aligned} \tag{1.6}$$

This system is solved using a novel combination of symbolic and numerical computer programming that provides an analytical expression for the displacement of a beam with two arbitrarily positioned rigid supports, one in each segment. For the method used here, this appears to be a new result.

System 3 The governing equation for each beam in a three-beam frame in terms of dimensional quantities is

$$E_j I_j \frac{d^4 y_j}{dx_j^4} = g_j(x_j) + f_{c,j}\delta(x_j - L_{c,j}) \quad j = 1, 2, 3 \quad \text{N/m} \tag{1.7}$$

where E_j is the tensile modulus of the *j*th beam and I_j is the moment of inertia of the cross section of the *j*th beam. We see from Eqs. (1.1) to (1.3) that the solutions to Eq. (1.7) are

$$\begin{aligned} y_j(x_j) &= d'(x_j)y_j(0) + d(x_j)y_j'(0) + v'(x_j, 0)y_j''(0) \\ &\quad + v(x_j, 0)y_j'''(0) + H_j(x_j) \qquad \text{m} \end{aligned} \tag{1.8}$$

where

$$H_j(x_j) = \frac{1}{E_j I_j}\left(h_j(x_j) + F_{c,j}\Phi(x_j, L_{c,j})\right) \qquad \text{m} \tag{1.9}$$

There are 12 solution constants that have to be determined: Six will be from the boundary conditions, and six will be from the continuity conditions where each pair of beams meets. It will turn out that five of these solution constants immediately can be set to zero, leaving a system of seven unknown solution constants.

This solution method provides the displacement of the entire frame, the moment and shear forces at the ends of each frame member, and the moment distribution in each frame member. It appears that obtaining the displacement response for the entire frame in this manner is a new result.

System 4 For a plate hinged on two opposite edges, one edge at the nondimensional location $\xi = 0$ and the other at $\xi = \beta_o$, we assume a separable solution of the form

$$\hat{w}(\eta, \xi) = \sum_{m=1}^{\infty} X_m(\eta)\sin(\gamma_m \xi) \tag{1.10}$$

where $\hat{w}$ is the nondimensional displacement of the plate and $\gamma_m = m\pi/\beta_o$. Upon substituting Eq. (1.10) into the governing plate equation and performing some additional operations, we arrive at

$$\frac{d^4X_m}{d\eta^4} + u_o\frac{d^2X_m}{d\eta^2} + u_fX_m = g_m\hat{f}_\eta(\eta) \tag{1.11}$$

where u_o represents an in-plane force in the η-direction and u_f represents the effects of the elastic foundation and an in-plane force in the ξ-direction. The quantity g_m represents the mth component of the magnitude of the force in the ξ-direction, and $\hat{f}_\eta(\eta)$ is the variation of the distributed load in the η-direction.

We see from (1.1) to (1.3) that the solution to Eq. (1.11) is

$$\begin{aligned} X_m(\eta) &= d_p'(\eta)X_m(0) + d_p(\eta)X_m'(0) + v_p'(\eta,0)X_m''(0) \\ &\quad + v_p(\eta,0)X_m'''(0) + g_mh(\eta) \end{aligned} \tag{1.12}$$

where d_p and v_p are solution functions, the prime denotes the derivative with respect to η, and h is the solution from the application of a distributed force. Upon comparing Eqs. (1.11) to (1.1), we see that S_o can be replaced with u_o and K_f with u_f. Thus, the results for system 1 can be used directly.

System 5 The governing equations for a circular arch in terms of nondimensional quantities are

$$\begin{aligned} &(1+\lambda)\frac{d^2v}{d\varphi^2} + \lambda\frac{d^3v}{d\varphi^3} - \frac{dw}{d\varphi} + \hat{Q}_\varphi(\varphi) = 0 \\ &\frac{dv}{d\varphi} - w - \lambda\left(\frac{d^4w}{d\varphi^4} + \frac{d^3v}{d\varphi^3}\right) + \hat{Q}_r(\varphi) = 0 \end{aligned} \tag{1.13}$$

where v is the nondimensional displacement in the tangential direction, w is the nondimensional displacement in the radial direction, λ is the ratio of the square of the radius of gyration of the cross section to the square of the radius of the circular arch, φ is the angular coordinate, $\hat{Q}_\varphi(\varphi)$ is the tangential component of the applied forces, and $\hat{Q}_r(\varphi)$ is the radial component of the applied forces.

Using the same Laplace transform procedure to obtain the solution to Eq. (1.1), the solutions to Eq. (1.13) are

$$\begin{aligned} w(\varphi) &= g_1(\varphi)V_0 + g_2(\varphi)v'(0) \\ &\quad + g_4'(\varphi)w''(0) + g_4''(\varphi)w'(0) \\ &\quad + (1-g_2(\varphi))\,w(0) + \hat{H}_r(\varphi)) \end{aligned} \tag{1.14}$$

and

$$
\begin{aligned}
v(\varphi) = &\left(-g_1'(\varphi) + g_4'(\varphi)\right) V_0 + g_3(\varphi)v'(0) \\
&+ g_4(\varphi)w''(0) + g_4'(\varphi)w'(0) \\
&+ (\varphi - g_3(\varphi))\, w(0) + v(0) + \hat{H}_\varphi(\varphi)
\end{aligned}
\tag{1.15}
$$

where $0 \leq \varphi \leq \beta_a$, $g_j(\varphi)$ are the solution functions, $\hat{H}_r(\varphi)$ is the solution from the application of the radial components of the applied forces, and $\hat{H}_\varphi(\varphi)$ is the solution from the application of the tangential components of the applied forces.

Equations (1.14) and (1.15) simplify as follows. For the arch clamped at $\varphi = 0$, we set $w(0) = v(0) = w'(0) = 0$, and for the arch hinged at $\varphi = 0$, we set $w(0) = v(0) = 0$ and $w''(0) = -v'(0)$. The remaining solution constants are determined from the boundary conditions at $\varphi = \beta_a$.

References

1. Kassimali A (2011) Structural analysis, 4th edn. Cengage Learning, Stamford
2. Udoeyo FF (2020) Structural analysis. North Broad Press an Imprint of Temple
3. Budynas RG, Sadegh AM (2020) Roark's formulas for stress and strain, 9th edn. McGraw-Hill/University Press, New York/Philadelphia
4. Fares N (2012) Practical approximate analysis of beams and frames. American Society of Civil Engineers, Reston
5. Karnovsky A, Lebed O (2010) Advanced methods of structural analysis. Springer, New York
6. Leontovich V (1959) Frames and arches. McGraw-Hill, New York
7. Pilkey WD (2002) Analysis and design of elastic beams. In: Computational methods. Wiley, New York
8. (2023) Manual of steel construction, 16th edn. American Institute of Steel Construction, Chicago

Uniform Beams

2

Abstract

Expressions for the spatial distribution of the displacement, rotation, moment, and shear force responses of a constant cross-section beam that is simultaneously subjected to a compressive axial force, concentrated forces, distributed forces, and applied moments and is resting on an elastic foundation and on in-span rigid supports are obtained for four sets of boundary conditions: clamped-clamped, clamped-free, clamped-hinged, and hinged-hinged. In addition, for any combination of these conditions, the lowest buckling load is determined. The solution is obtained by using the Laplace transform on the spatial variable and by creating a special set of solution functions. This solution will then be used in subsequent chapters to obtain the response of beams with a step change in flexural rigidity, three-beam, five-beam, and six-beam rectangular frames, and a rectangular plate with two opposite edges hinged.

Keywords

Beams · Axial force · Elastic foundation · Rigid supports · Concentrated force · Distributed force · Moments · Boundary conditions · Hinged · Clamped · Free · Buckling · Displacement · Shear force

---

Supplementary Information The online version contains supplementary material available at (https://doi.org/10.1007/978-3-032-28726-7_2).

E. B. Magrab, *Static Analysis of Beams, Frames, and Arches*, Synthesis Lectures on Mechanical Engineering, https://doi.org/10.1007/978-3-032-28726-7_2

2.1 Governing Equations

Consider a beam of constant cross section, length L (m), tensile modulus E (N/m^2), and cross-section moment of inertia I (m^4). As shown in Fig. 2.1a, the beam is restrained by a linear elastic spring k_1 (N/m) at $x = L_1, 0 \leq L_1 \leq L$, and another linear elastic spring k_2 (N/m) at $x = L_2, 0 \leq L_1 < L_2 \leq L$. The introduction of these in-span springs will be used at an appropriate point in the solution process to obtain expressions for two rigid in-span supports by letting $k_j \to \infty$. This method has been used by the author to determine the natural frequencies of beams with in-span rigid supports [1, 2]. In addition, an axial compressive load P_o (N) is applied, and the entire beam rests on an elastic foundation of stiffness k_f (N/m^2). Referring to Figs. 2.1b, 2.1c, and 2.1d, respectively, the beam is subject to three independently applied nonoverlapping trapezoidal distributed forces $h_j(x)$ (N/m), j = 1, 2, 3, three independently applied concentrated forces $f_{c,j}$ (N) at $x = L_{c,j}$ (m), $0 \leq L_{c,j} \leq L$, and three independently applied counterclockwise moments t_j (Nm) at $x = L_{T,j}$ (m), $0 \leq L_{T,j} \leq L$.

If the displacement of the beam is $y = y(x)$ (m), then the governing equation for the beam can be written as

$$\begin{aligned} EI\frac{d^4y}{dx^4} + P_o\frac{d^2y}{dx^2} + k_f y + \sum_{j=1}^{2} k_j\delta(x - L_j)y &= \sum_{j=1}^{3} h_j(x) \\ + \sum_{j=1}^{3} f_{c,j}\delta(x - L_{c,j}) &- \sum_{j=1}^{3} t_j\frac{d}{dx}\delta(x - L_{T,j}) \quad \text{N/m} \end{aligned} \tag{2.1}$$

where $\delta(x)$ is the delta function. We assume that $h_j(x)$ has the trapezoidal form

$$h_j(x) = (a_j x + b_j)\big(u(x - L_{L,j}) - u(x - L_{R,j})\big) \tag{2.2}$$

where $u(x)$ is the unit step function and

$$\begin{aligned} a_j &= \frac{F_{L,j} - F_{R,j}}{L_{L,j} - L_{R,j}} \quad \text{N/m}^2 \\ b_j &= \frac{F_{R,j}L_{L,j} - F_{L,j}L_{R,j}}{L_{L,j} - L_{R,j}} \quad \text{N/m} \end{aligned} \tag{2.3}$$

In Eq. (2.3), $F_{L,j}$ (N/m) and $F_{R,j}$ (N/m), respectively, are the magnitudes of the end points of the trapezoidal load located at $L_{L,j}$ and $L_{R,j}$, where $0 \leq L_{L,1} < L_{R,1} \leq L_{L,2} < L_{R,2} \leq L_{L,3} < L_{R,3} \leq L$. It is noted that the slope of the beam ϕ is $\phi = dy/dx$, the moment m (Nm) is $m = EId^2y/dx^2$, and the shear force v_s (N) is $v_s = EId^3y/dx^3$.

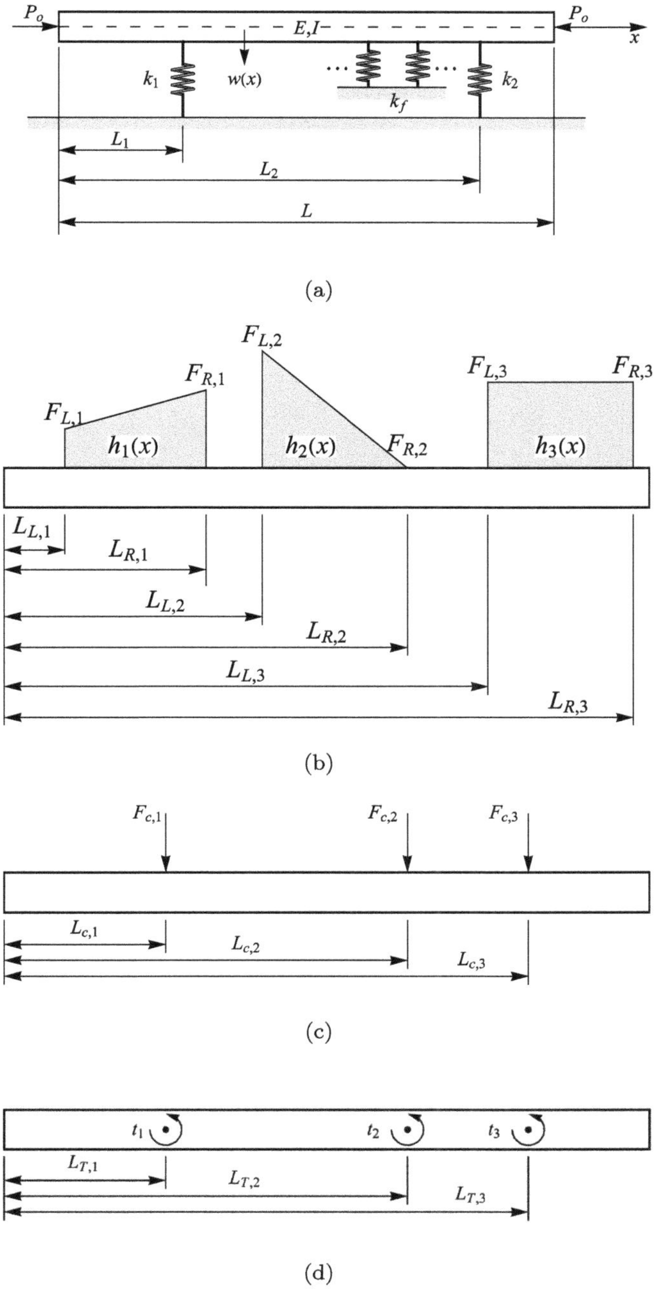

Fig. 2.1 (**a**) Nomenclature for a beam resting on an elastic foundation, subjected to a compressive axial load, and has attached two in-span linear springs; (**b**) nomenclature for the trapezoidal distributed forces; (**c**) nomenclature for the concentrated forces; and (**d**) nomenclature for the counterclockwise moments

To place Eqs. (2.1)–(2.3) in a more useful form, we introduce the following nondimensional quantities $\eta = x/L$ and

$$w = \frac{EIy}{F_{max}L^3} \tag{2.4}$$

where F_{max} (N) will be defined subsequently. Upon substituting these definitions into Eq. (2.1), we arrive at

$$\begin{aligned} \frac{d^4w}{d\eta^4} + S_o\frac{d^2w}{d\eta^2} + K_f w + \sum_{j=1}^{2} K_j\delta(\eta - \eta_j)w &= \sum_{j=1}^{3} g_j(\eta) \\ &+ \sum_{j=1}^{3} F_{c,j}\delta(\eta - \eta_{c,j}) - \sum_{j=1}^{3} T_j\frac{d}{d\eta}\delta(\eta - \eta_{T,j}) \end{aligned} \tag{2.5}$$

where $\eta_j = L_j/L, \eta_{c,j} = L_{c,j}/L, \eta_{T,j} = L_{T,j}/L,$

$$\begin{aligned} S_o &= \frac{P_oL^2}{EI} \qquad K_j = \frac{k_jL^3}{EI} \qquad K_f = \frac{k_fL^4}{EI} \\ F_{c,j} &= \frac{f_{c,j}}{F_{max}} \qquad T_j = \frac{t_j}{F_{max}L} \end{aligned} \tag{2.6}$$

are nondimensional quantities,

$$g_j(\eta) = (\hat{a}_j\eta + \hat{b}_j)\big(u(\eta - \eta_{L,j}) - u(\eta - \eta_{R,j})\big) \tag{2.7}$$

where $\eta_{L,j} = L_{L,j}/L,\ \eta_{R,j} = L_{R,j}/L$, and

$$\begin{aligned} \hat{a}_j &= \frac{q_{L,j} - q_{R,j}}{\eta_{L,j} - \eta_{R,j}} \\ \hat{b}_j &= \frac{q_{R,j}\eta_{L,j} - q_{L,j}\eta_{R,j}}{\eta_{L,j} - \eta_{R,j}} \end{aligned} \tag{2.8}$$

are also nondimensional quantities. In Eq. (2.8), $q_{L,j} = LF_{L,j}/F_{max}$ and $q_{R,j} = LF_{R,j}/F_{max}$. It is seen from Eq. (2.6) that $S_o = K_f$ when $P_o = k_fL^2$ (N). It is mentioned that $g_j(\eta)$ can be combined to form triangular and more general trapezoidal distributed loads.

The quantity F_{max} is selected as the maximum value of the following quantities: $F_{c,j}$ (N), t_j/L (N), $LF_{L,j}$ (N), and $LF_{R,j}$ (N). Therefore, depending on which entities are selected, at least one of the quantities will have the value of 1 and the rest of them will have a value that is less than or equal to 1. Thus, when more than one of these loads exists simultaneously, one must have an estimate of their respective magnitudes and the length of the beam.

Table 2.1 Beam boundary conditions, where the prime (′) is the derivative with respect to η

Case	Boundary conditions	$\eta = 0$	$\eta = 1$
1	Clamped-clamped	$w(0) = 0$	$w(1) = 0$
		$w'(0) = 0$	$w'(1) = 0$
2	Clamped-free	$w(0) = 0$	$w''(1) = 0$
		$w'(0) = 0$	$w'''(1) + S_o w'(1) = 0$
3	Clamped-hinged	$w(0) = 0$	$w(1) = 0$
		$w'(0) = 0$	$w''(1) = 0$
4	Hinged-hinged	$w(0) = 0$	$w(1) = 0$
		$w''(0) = 0$	$w''(1) = 0$

In terms of the nondimensional quantities, we find that the slope of the beam ϕ becomes

$$\Phi = \frac{EI\phi}{F_{max}L^2} = \frac{dw}{d\eta} \tag{2.9}$$

the moment m becomes

$$M = \frac{m}{F_{max}L} = \frac{d^2w}{d\eta^2} \tag{2.10}$$

and the shear force v_s becomes

$$V = \frac{v_s}{F_{max}} = \frac{d^3w}{d\eta^3} \tag{2.11}$$

These are the quantities for which numerical values are determined. These relations are also used to convert the numerical values into dimensional quantities.

The boundary conditions considered are given in Table 2.1 and are represented graphically in Fig. 2.2 for the cases where the axial force is present or absent. Other combinations of boundary conditions can also be considered with this method, such as attaching a torsion spring to a hinged end.

The solution to Eq. (2.5) with the boundary conditions given in Table 2.1 is obtained using the Laplace transform, which easily handles the generalized functions: the delta function and the unit step function.[1] If s is the Laplace transform variable and $\bar{w}(s)$ is the Laplace transform of $w(\eta)$, then taking the Laplace transform of Eq. (2.5) with respect to

[1] The author first became aware in the 1960s of the power of this approach to solve boundary value problems from the translated work of W. Nowacki 1963 Dynamics of Elastic Systems, John Wiley & Sons, New York, who applied the Laplace transform to the spatial variable to determine the response of a number of different vibrating elastic systems.

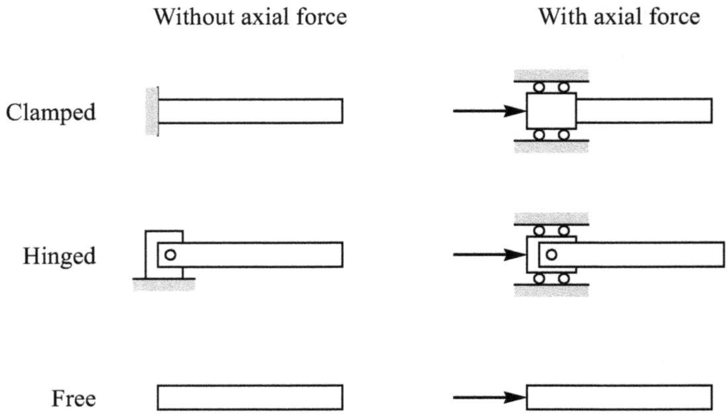

Fig. 2.2 Typical representation of the boundary conditions

η, solving for $\bar{w}(s)$, and then taking the inverse Laplace transform yield[2]

$$w(\eta) = d'(\eta)w(0) + d(\eta)w'(0) + v'(\eta, 0)w''(0) + v(\eta, 0)w'''(0) + H(\eta) - K(\eta) \tag{2.12}$$

where $w(0)$, $w'(0)$, $w''(0)$, and $w'''(0)$ are the solution constants that are determined from the specification of the boundary conditions,

$$\begin{aligned} H(\eta) &= \sum_{j=1}^{3} \left(\hat{a}_j A_j(\eta) + \hat{b}_j B_j(\eta) \right) + \sum_{j=1}^{3} F_{c,j} v(\eta, \eta_{c,j}) \theta(\eta - \eta_{c,j}) \\ &\quad - \sum_{j=1}^{3} T_j v'(\eta, \eta_{T,j}) \theta(\eta - \eta_{T,j}) \\ K(\eta) &= \sum_{j=1}^{2} K_j w(\eta_j) v(\eta, \eta_j) \theta(\eta - \eta_j) \end{aligned} \tag{2.13}$$

and

$$\begin{aligned} A_j(\eta) &= \left(\eta_{L,j} r'(\eta, \eta_{L,j}) + r(\eta, \eta_{L,j})\right) \theta(\eta - \eta_{L,j}) \\ &\quad - \left(\eta_{R,j} r'(\eta, \eta_{R,j}) + r(\eta, \eta_{R,j})\right) \theta(\eta - \eta_{R,j}) \\ B_j(\eta) &= r'(\eta, \eta_{L,j}) \theta(\eta - \eta_{L,j}) - r'(\eta, \eta_{R,j}) \theta(\eta - \eta_{R,j}) \end{aligned} \tag{2.14}$$

[2] These operations were performed using Mathematica®.

where $\theta(x)$ is the Heaviside theta function, which is defined as equal to 1 when $x > 0$ and equal to 0 when $x < 0$. It is not defined at $x = 0$.

The nondimensional solution functions $r(\eta, \alpha)$, $v(\eta, \alpha)$, and $d(\eta)$, respectively, are given in Tables 2.2, 2.3, and 2.4 for four scenarios: axial force and elastic foundation, axial force only, elastic foundation only, and no axial force and no elastic foundation. By introducing these functions in the manner shown, one is able to obtain the solutions for these four cases without explicitly specifying each case with a separate solution. Furthermore, these functions are a shorthand notation for the more explicit notation

$$\begin{aligned} r(\eta, \alpha) &= r(\eta, \alpha, S_o, K_f) \\ v(\eta, \alpha) &= v(\eta, \alpha, S_o, K_f) \\ d(\eta) &= d(\eta, S_o, K_f) \end{aligned} \tag{2.15}$$

and has been introduced to make the subsequent results more readable. It is mentioned that $v(\eta, \alpha)$ and $d(\eta)$ are solutions to the homogeneous differential equation and that the function $r(\eta, \alpha)$ is a function that describes the solution for the distributed load.

Solutions' Region of Validity Consider a beam configuration for which one has selected a numerical value for K_f and the in-span spring locations. Let the lowest buckling load for this configuration be $\hat{S}_{o,1}$. Then the solution given by Eq. (2.12) will be valid only when $S_o < \hat{S}_{o,1}$. From a practical point of view, $S_o < 0.9\hat{S}_{o,1}$ should be used.

It is noted that all the terms appearing in Eqs. (2.13) and (2.14) are of the form $G(x, x_o) = g(x, x_o)\theta(x - x_o)$. The derivative of this product with respect to x is $G'(x, x_o) = g'(x, x_o)\theta(x - x_o) + g(x, x_o)\delta(x - x_o)$. However, in the second term, $\delta(x - x_o)$ is zero everywhere except at $x = x_o$. Therefore, if we restrict the values of x to those for which $x > x_o > 0$, then $G'(x, x_o) = g'(x, x_o)$, and, in general, it can be shown that $G^{(n)}(x, x_o) = g^{(n)}(x, x_o)$ for $x > x_o > 0$ and equals 0 for $x < x_o$. Thus, in Eq. (2.13), $w(\eta_j)$ is shorthand for $w(\eta_j + \epsilon)$, where $0 < \epsilon << 1$.

As indicated in Sect. 1.2, in Eq. (2.12), $w(0)$ is the displacement at $\eta = 0$, $w'(0)$ is the slope at $\eta = 0$, $w''(0)$ is proportional to the moment at $\eta = 0$, and $w'''(0)$ is proportional to the shear force at $\eta = 0$. These quantities are determined from the boundary conditions given in Table 2.1. Therefore, from Table 2.1, Eq. (2.12) can be simplified as follows. For cases 1, 2, and 3, Eq. (2.12) reduces to

$$w(\eta) = v'(\eta, 0)w''(0) + v(\eta, 0)w'''(0) + H(\eta) - K(\eta) \tag{2.16}$$

and for case 4, Eq. (2.12) reduces to

$$w(\eta) = d(\eta)w'(0) + v(\eta, 0)w'''(0) + H(\eta) - K(\eta) \tag{2.17}$$

Table 2.2 The nondimensional solution function $r(\eta, \alpha)$ and its derivatives for four combinations of factors: axial force and elastic foundation, axial force only, elastic foundation only, and none of these. The prime ($'$) denotes the derivative with respect to η, $b = 0.5\sqrt{S_o^2 - 4K_f}$, $a = -S_o/2$, $\gamma_{\eta\alpha} = (\eta - \alpha)\sqrt[4]{K_f}/\sqrt{2}$, $\varepsilon_{\eta\alpha} = (\eta - \alpha)\sqrt{-a-b}$, $\kappa_{\eta\alpha} = (\eta - \alpha)\sqrt{b-a}$, and $\lambda_{\eta\alpha} = (\eta - \alpha)\sqrt{S_o}$

$r^{(n)}(\eta, \alpha)$	$S_o \neq 0$ and $K_f > 0$,	$S_o \neq 0$ and $K_f = 0$	$S_o = 0$ and $K_f > 0$,	$S_o = 0$ and $K_f = 0$
$r(\eta, \alpha)$	$\frac{\eta-\alpha}{a^2-b^2} - \frac{\sin(\varepsilon_{\eta\alpha})}{2b(-a-b)^{3/2}}$ $+\frac{\sin(\kappa_{\eta\alpha})}{2b(b-a)^{3/2}}$	$-\frac{\eta-\alpha}{S_o^2} + \frac{(\eta-\alpha)^3}{6S_o} + \frac{1}{S_o^{5/2}}\sin\left(\lambda_{\eta\alpha}\right)$	$\frac{1}{K_f}(\eta - \alpha) - \frac{1}{\sqrt{2}K_f^{5/4}}$ $\Big(\cosh(\gamma_{\eta\alpha})\sin(\gamma_{\eta\alpha})$ $+\cos(\gamma_{\eta\alpha})\sinh(\gamma_{\eta\alpha})\Big)$	$\frac{1}{120}(\eta - \alpha)^5$
$r'(\eta, \alpha)$	$\frac{1}{a^2-b^2} + \frac{\cos(\varepsilon_{\eta\alpha})}{2b(a+b)} + \frac{\cos(\kappa_{\eta\alpha})}{2b(b-a)}$	$\frac{(\eta-\alpha)^2}{2S_o} - \frac{1}{S_o^2}\left(1 - \cos(\lambda_{\eta\alpha})\right)$	$\frac{1}{K_f} - \frac{1}{K_f}\cos(\gamma_{\eta\alpha})\cosh(\gamma_{\eta\alpha})$	$\frac{1}{24}(\eta - \alpha)^4$
$r''(\eta, \alpha)$	$\frac{\sin(\varepsilon_{\eta\alpha})}{2b\sqrt{-a-b}} - \frac{\sin(\kappa_{\eta\alpha})}{2b\sqrt{b-a}}$	$\frac{\eta-\alpha}{S_o} - \frac{1}{S_o^{3/2}}\sin(\lambda_{\eta\alpha})$	$\frac{1}{\sqrt{2}K_f^{3/4}}$ $\Big(\cosh(\gamma_{\eta\alpha})\sin(\gamma_{\eta\alpha})$ $-\cos(\gamma_{\eta\alpha})\sinh(\gamma_{\eta\alpha})\Big)$	$\frac{1}{6}(\eta - \alpha)^3$
$r'''(\eta, \alpha)$	$\frac{1}{2b}\left(\cos(\varepsilon_{\eta\alpha}) - \cos(\kappa_{\eta\alpha})\right)$	$\frac{1}{S_o}\left(1 - \cos(\lambda_{\eta\alpha})\right)$	$\frac{1}{\sqrt{K_f}}\sin(\gamma_{\eta\alpha})\sinh(\gamma_{\eta\alpha})$	$\frac{1}{2}(\eta - \alpha)^2$
$r''''(\eta, \alpha)$	$\frac{1}{2b}\Big(-\sqrt{-a-b}\sin(\varepsilon_{\eta\alpha})$ $+\sqrt{b-a}\sin(\kappa_{\eta\alpha})\Big)$	$\frac{1}{\sqrt{S_o}}\sin(\lambda_{\eta\alpha})$	$\frac{1}{\sqrt{2}\sqrt[4]{K_f}}\Big(\cosh(\gamma_{\eta\alpha})\sin(\gamma_{\eta\alpha})$ $+\cos(\gamma_{\eta\alpha})\sinh(\gamma_{\eta\alpha})\Big)$	$\eta - \alpha$

Table 2.3 The nondimensional solution function $v(\eta, \alpha)$ and its derivatives for four combinations of factors: axial force and elastic foundation, axial force only, elastic foundation only, and none of these. The prime ($'$) denotes the derivative with respect to η, $b = 0.5\sqrt{S_o^2 - 4K_f}$, $a = -S_o/2$, $\gamma_{\eta\alpha} = (\eta - \alpha)\sqrt[4]{K_f}/\sqrt{2}$, $\varepsilon_{\eta\alpha} = (\eta - \alpha)\sqrt{-a-b}$, $\kappa_{\eta\alpha} = (\eta - \alpha)\sqrt{b-a}$, and $\lambda_{\eta\alpha} = (\eta - \alpha)\sqrt{S_o}$

$v^{(n)}(\eta, \alpha)$	$S_o \neq 0$ and $K_f > 0$,	$S_o \neq 0$ and $K_f = 0$	$S_o = 0$ and $K_f > 0$,	$S_o = 0$ and $K_f = 0$
$v(\eta, \alpha)$	$\frac{\sin(\varepsilon_{\eta\alpha})}{2b\sqrt{-a-b}} - \frac{\sin(\kappa_{\eta\alpha})}{2b\sqrt{b-a}}$	$\frac{\eta-\alpha}{S_o} - \frac{1}{S_o^{3/2}}\sin(\lambda_{\eta\alpha})$	$\frac{1}{\sqrt{2}K_f^{3/4}}\big(\cosh(\gamma_{\eta\alpha})\sin(\gamma_{\eta\alpha})$ $-\cos(\gamma_{\eta\alpha})\sinh(\gamma_{\eta\alpha})\big)$	$\frac{1}{6}(\eta-\alpha)^3$
$v'(\eta, \alpha)$	$\frac{1}{2b}\big(\cos(\varepsilon_{\eta\alpha}) - \cos(\kappa_{\eta\alpha})\big)$	$\frac{1}{S_o}\big(1 - \cos(\lambda_{\eta\alpha})\big)$	$\frac{1}{\sqrt{K_f}}\sin(\gamma_{\eta\alpha})\sinh(\gamma_{\eta\alpha})$	$\frac{1}{2}(\eta-\alpha)^2$
$v''(\eta, \alpha)$	$-\frac{\sqrt{-a-b}}{2b}\sin(\varepsilon_{\eta\alpha}) + \frac{\sqrt{b-a}}{2b}\sin(\kappa_{\eta\alpha})$	$\frac{1}{\sqrt{S_o}}\sin(\lambda_{\eta\alpha})$	$\frac{1}{\sqrt{2}\sqrt[4]{K_f}}\big(\cosh(\gamma_{\eta\alpha})\sin(\gamma_{\eta\alpha})$ $+\cos(\gamma_{\eta\alpha})\sinh(\gamma_{\eta\alpha})\big)$	$\eta - \alpha$
$v'''(\eta, \alpha)$	$\frac{a+b}{2b}\cos(\varepsilon_{\eta\alpha}) + \frac{b-a}{2b}\cos(\kappa_{\eta\alpha})$	$\cos(\lambda_{\eta\alpha})$	$\cosh(\gamma_{\eta\alpha})\cos(\gamma_{\eta\alpha})$	1
$v''''(\eta, \alpha)$	$\frac{(-a-b)^{3/2}}{2b}\sin(\varepsilon_{\eta\alpha}) - \frac{(b-a)^{3/2}}{2b}\sin(\kappa_{\eta\alpha})$	$-\sqrt{S_o}\sin(\lambda_{\eta\alpha})$	$\sqrt[4]{K_f}\big(-\cosh(\gamma_{\eta\alpha})\sin(\gamma_{\eta\alpha})$ $+\cos(\gamma_{\eta\alpha})\sinh(\gamma_{\eta\alpha})\big)/\sqrt{2}$	0

Table 2.4 The nondimensional solution function $d(\eta)$ and its derivatives for four combinations of factors: axial force and elastic foundation, axial force only, elastic foundation only, and none of these. The prime (′) denotes the derivative with respect to η, $b = 0.5\sqrt{S_o^2 - 4K_f}$, $a = -S_o/2$, $\varepsilon_\eta = \eta\sqrt{-a-b}$, $\kappa_\eta = \eta\sqrt{b-a}$, and $\lambda_\eta = \eta\sqrt[4]{K_f}/\sqrt{2}$

$d^{(n)}(\eta)$	$S_o \neq 0$ and $K_f > 0$,	$S_o \neq 0$ and $K_f = 0$	$S_o = 0$ and $K_f > 0$,	$S_o = 0$ and $K_f = 0$
$d(\eta)$	$S_o v(\eta, 0)$ $-\frac{1}{2b}\big(\sqrt{-a-b}\sin(\varepsilon_\eta)$ $-\sqrt{b-a}\sin(\kappa_\eta)\big)$	η	$\frac{1}{\sqrt{2}\sqrt[4]{K_f}}$ $\big(\cosh(\lambda_\eta)\sin(\lambda_\eta)$ $+\cos(\lambda_\eta)\sinh(\lambda_\eta)\big)$	η
$d'(\eta)$	$S_o v'(\eta, 0)$ $+\frac{1}{2b}\big((a+b)\cos(\varepsilon_\eta)$ $+(b-a)\cos(\kappa_\eta)\big)$	1	$\cosh(\lambda_\eta)\cos(\lambda_\eta)$	1
$d''(\eta)$	$S_o v''(\eta, 0)$ $+\frac{1}{2b}\big((-a-b)^{3/2}\sin(\varepsilon_\eta)$ $-(b-a)^{3/2}\sin(\kappa_\eta)\big)$	0	$\sqrt[4]{K_f}$ $\big(-\cosh(\lambda_\eta)\sin(\lambda_\eta)$ $+\cos(\lambda_\eta)\sinh(\lambda_\eta)\big)/\sqrt{2}$	0
$d'''(\eta)$	$S_o v'''(\eta, 0)$ $+\frac{1}{2b}\big((b+a)^2\cos(\varepsilon_\eta)$ $-(a-b)^2\cos(\kappa_\eta)\big)$	0	$-\sqrt{K_f}\sin(\lambda_\eta)\sinh(\lambda_\eta)$	0
$d''''(\eta)$	$S_o v''''(\eta, 0)$ $-\frac{1}{2b}\big((-a-b)^{5/2}\sin(\varepsilon_\eta)$ $-(b-a)^{5/2}\sin(\kappa_\eta)\big)$	0	$-K_f^{3/4}$ $\big(\cosh(\lambda_\eta)\sin(\lambda_\eta)$ $+\cos(\lambda_\eta)\sinh(\lambda_\eta)\big)/\sqrt{2}$	0

The remaining unknown solution constants, $w''(0)$ and $w'''(0)$ in Eq. (2.16) and $w'(0)$ and $w'''(0)$ in Eq. (2.17), are determined from the boundary conditions at $\eta = 1$. Thus, another advantage of the Laplace transform method is that in certain circumstances one can immediately eliminate two of the four solution constants, which, as will be shown, makes the subsequent algebraic manipulations feasible. It is seen from Eqs. (2.16), (2.17), and (2.13) that there are two additional unknown quantities to be determined: $w(\eta_1)$ and $w(\eta_2)$. Their determination will be discussed subsequently when we obtain the solution for the rigid in-span supports.

Using the appropriate equation, either Eq. (2.16) or Eq. (2.17), in cases 1 to 4 in Table 2.1, we obtain the results presented in the subsequent sections.

2.2 Clamped-Clamped Beam

For a beam clamped at both ends, we use Eq. (2.16) and the boundary conditions at $\eta = 1$ given in Table 2.1 for case 1. This yields the following system of equations:

$$\begin{aligned} w(1) &= v'(1,0)w''(0) + v(1,0)w'''(0) + H(1) - K(1) = 0 \\ w'(1) &= v''(1,0)w''(0) + v'(1,0)w'''(0) + H'(1) - K'(1) = 0 \end{aligned} \tag{2.18}$$

where

$$\begin{aligned} K(1) &= K_1 w(\eta_1)v(1,\eta_1) + K_2 w(\eta_2)v(1,\eta_2) \\ K'(1) &= K_1 w(\eta_1)v'(1,\eta_1) + K_2 w(\eta_2)v'(1,\eta_2) \end{aligned} \tag{2.19}$$

Solving Eq. (2.18) for $w''(0)$ and $w'''(0)$, we obtain

$$\begin{aligned} w''(0) &= \frac{1}{D_{cc}}\Big(v'(1,0)\big(-H(1)+K(1)\big) + v(1,0)\big(H'(1)-K'(1)\big)\Big) \\ w'''(0) &= \frac{1}{D_{cc}}\Big(v''(1,0)\big(H(1)-K(1)\big) + v'(1,0)\big(-H'(1)+K'(1)\big)\Big) \end{aligned} \tag{2.20}$$

where

$$D_{cc} = [v'(1,0)]^2 - v(1,0)v''(1,0) \tag{2.21}$$

is the determinant of the coefficients of Eq. (2.18).

Substituting Eq. (2.20) into Eq. (2.16), using Eq. (2.19), and rearranging terms give

$$w_{cc}(\eta) = \frac{1}{D_{cc}}\left(R_{cc}(\eta) + K_1 w_{cc}(\eta_1)S_{cc}(\eta,\eta_1) + K_2 w_{cc}(\eta_2)S_{cc}(\eta,\eta_2)\right) \tag{2.22}$$

where

$$\begin{aligned}
R_{cc}(\eta) &= D_{cc}H(\eta) + P_{cc}(\eta)\\
P_{cc}(\eta) &= v'(\eta,0)\big(v(1,0)H'(1) - v'(1,0)H(1)\big)\\
&\quad - v(\eta,0)\big(v'(1,0)H'(1) - v''(1,0)H(1)\big)\\
S_{cc}(\eta,\eta_j) &= -D_{cc}v(\eta,\eta_j)u(\eta-\eta_j)\\
&\quad + v'(\eta,0)\big(v(1,\eta_j)v'(1,0) - v(1,0)v'(1,\eta_j)\big)\\
&\quad + v(\eta,0)\big(v'(1,0)v'(1,\eta_j) - v(1,\eta_j)v''(1,0)\big)
\end{aligned} \tag{2.23}$$

No In-span Rigid Supports For the case of no in-span rigid supports, we set $K_1 = K_2 = 0$, and Eq. (2.22) simplifies to

$$w_{cc}(\eta) = \frac{R_{cc}(\eta)}{D_{cc}} \tag{2.24}$$

One Rigid In-span Support To obtain the solution for the case of one rigid in-span support, we set $K_2 = 0$ in Eq. (2.22), Eq. (2.22) becomes

$$w_{cc}(\eta) = \frac{1}{D_{cc}}\big(R_{cc}(\eta) + K_1 w_{cc}(\eta_1)S_{cc}(\eta,\eta_1)\big) \tag{2.25}$$

Setting $\eta = \eta_1$ in Eq. (2.25) and solving for $w_{cc}(\eta_1)$ yield

$$w_{cc}(\eta_1) = \frac{R_{cc}(\eta_1)}{D_{cc} - K_1 S_{cc}(\eta_1,\eta_1)r} \tag{2.26}$$

Substituting Eq. (2.26) into Eq. (2.25), we obtain

$$w_{cc}(\eta) = \frac{1}{D_{cc}}\left(R_{cc}(\eta) + \frac{K_1 R_{cc}(\eta_1)}{D_{cc} - K_1 S_{cc}(\eta_1,\eta_1)}S_{cc}(\eta,\eta_1)\right) \tag{2.27}$$

Taking the limit in Eq. (2.27) as $K_1 \to \infty$ gives

$$w_{cc}(\eta) = \frac{1}{D_{cc}S_{cc}(\eta_1,\eta_1)}\big(R_{cc}(\eta)S_{cc}(\eta_1,\eta_1) - R_{cc}(\eta_1)S_{cc}(\eta,\eta_1)\big) \tag{2.28}$$

which is the displacement of a beam with one in-span rigid support placed at η_1. To verify this, we note that when $\eta = \eta_1$ in Eq. (2.28), $w_{cc}(\eta_1) = 0$. It is seen that after the limiting process, the determinant of the coefficients of the system of equations is $D_{cc}S_{cc}(\eta_1,\eta_1)$.

Properties of the Rigid In-span Supports The rigid in-span support is equivalent to placing a hinge support at its location, for it sets the displacement to zero but does not impede the beam from rotating and does not resist any internal moments.

Two Rigid In-span Supports To obtain the solution for the case of two rigid supports, we use Eq. (2.22). Upon setting $\eta = \eta_1$ in Eq. (2.22) and then setting $\eta = \eta_2$ in Eq. (2.22), we arrive at the following equations for $w_{cc}(\eta_1)$ and $w_{cc}(\eta_2)$

$$
\begin{aligned}
w_{cc}(\eta_1) &= \frac{-D_{cc}R_{cc}(\eta_1) + K_2E_{cc,1}(\eta_1,\eta_2)}{-D_{cc}^2 + K_1D_{cc}S_{cc}(\eta_1,\eta_1) + K_2D_{cc}S_{cc}(\eta_2,\eta_2) + K_1K_2E_{cc}(\eta_1,\eta_2)} \\
w_{cc}(\eta_2) &= \frac{-D_{cc}R_{cc}(\eta_2) + K_1E_{cc,2}(\eta_1,\eta_2)}{-D_{cc}^2 + K_1D_{cc}S_{cc}(\eta_1,\eta_1) + K_2D_{cc}S_{cc}(\eta_2,\eta_2) + K_1K_2E_{cc}(\eta_1,\eta_2)}
\end{aligned}
\tag{2.29}
$$

where

$$
\begin{aligned}
E_{cc}(\eta_1,\eta_2) &= S_{cc}(\eta_2,\eta_1)S_{cc}(\eta_1,\eta_2) - S_{cc}(\eta_1,\eta_1)S_{cc}(\eta_2,\eta_2) \\
&= D_{cc}Z_{cc}(\eta_1,\eta_2) \\
E_{cc,1}(\eta_1,\eta_2) &= -R_{cc}(\eta_2)S_{cc}(\eta_1,\eta_2) + R_{cc}(\eta_1)S_{cc}(\eta_2,\eta_2) \\
E_{cc,2}(\eta_1,\eta_2) &= R_{cc}(\eta_2)S_{cc}(\eta_1,\eta_1) - R_{cc}(\eta_1)S_{cc}(\eta_2,\eta_1)
\end{aligned}
\tag{2.30}
$$

and

$$
\begin{aligned}
Z_{cc}(\eta_1,\eta_2) = &\big(-v(1,\eta_1)v'(1,\eta_2) + v'(1,\eta_1)v(1,\eta_2)\big)\times \\
&\big(-v(\eta_1,0)v'(\eta_2,0) + v'(\eta_1,0)v(\eta_2,0)\big) \\
&+ v(\eta_2,\eta_1)\Big(v(1,0)v'(1,\eta_2)v(\eta_1,0) \\
&- v'(1,0)v(1,\eta_2)v'(\eta_1,0) - v'(1,0)v'(1,\eta_2)v(\eta_1,0) \\
&+ v''(1,0)v(1,\eta_2)v(\eta_1,0)\Big)
\end{aligned}
\tag{2.31}
$$

Substituting Eq. (2.29) into Eq. (2.22) and letting $K_1 \to \infty$ and $K_2 \to \infty$, we arrive at

$$
\begin{aligned}
w_{cc}(\eta) = &\frac{1}{D_{cc}E_{cc}(\eta_1,\eta_2)}\Big(R_{cc}(\eta)E_{cc}(\eta_1,\eta_2) + E_{cc,1}(\eta_1,\eta_2)S_{cc}(\eta,\eta_1) \\
&+ E_{cc,2}(\eta_1,\eta_2)S_{cc}(\eta,\eta_2)\Big)
\end{aligned}
\tag{2.32}
$$

which is the displacement of a beam with two in-span rigid supports, one placed at η_1 and the other at η_2. To verify this, it is straightforward to show from Eq. (2.32) that when

$\eta = \eta_1$, $w_{cc}(\eta_1) = 0$, and when $\eta = \eta_2$, $w_{cc}(\eta_2) = 0$. In addition, it is seen that the determinant of the coefficients of the system of equations to determine $w_{cc}(\eta_1)$ and $w_{cc}(\eta_2)$ is $D_{cc}E_{cc}(\eta_1, \eta_2)$.

2.2.1 Buckling Load

When S_o is compressive, the buckling loads can be obtained directly from Eqs. (2.24), (2.28), and (2.32) from their respective determinant of the coefficients used to satisfy the boundary conditions and the presence of the in-span rigid supports as follows. It is noted that this requires that the functions for r, v, and d be taken from Tables 2.2, 2.3, and 2.4, depending on whether or not $K_f = 0$. The buckling loads are determined by setting all the applied transverse forces and the moments to zero. This results in $R_{cc}(\eta) = 0$ and, consequently, $E_1 = E_2 = 0$.

Before setting these quantities to zero, we first multiply each of these three equations by the denominator appearing on the right-hand side of each equation to obtain the following. For no in-span rigid supports,

$$D_{cc}w_{cc}(\eta) = 0 \tag{2.33}$$

for one in-span rigid support at $\eta = \eta_1$,

$$D_{cc}S_{cc}(\eta_1, \eta_1)w_{cc}(\eta) = 0 \tag{2.34}$$

and for two in-span rigid supports, one at $\eta = \eta_1$ and the other at $\eta = \eta_2$,

$$D_{cc}E_{cc}(\eta_1, \eta_2)w_{cc}(\eta) = D_{cc}^2 Z_{cc}(\eta_1, \eta_2)w_{cc}(\eta) = 0 \tag{2.35}$$

Since $w_{cc}(\eta) = 0$ is a trivial solution for no in-span rigid supports, the buckling loads are the values of $S_o = S_{o,n}$ that satisfy

$$D_{cc} = 0 \tag{2.36}$$

where $n = 1$ is the lowest buckling load. For example, if $K_f = 0$, then using Eq. (2.21) and the functions appearing in the appropriate column of Table 2.3, we obtain

$$D_{cc} = 2\left(1 - \cos\sqrt{S_o}\right) - \sqrt{S_o}\sin\sqrt{S_o} = 0 \tag{2.37}$$

Evaluation of Eq. (2.37) gives the lowest buckling load $S_{o,1} = 39.48$.

It is seen from Eq. (2.15) that each function comprising D_{cc} is a function of S_o. Therefore, these functions are those for which $S_o > 0$ in Tables 2.2, 2.3, and 2.4. The same is true for the subsequent buckling load equations.

For one in-span rigid support, we note in Eq. (2.34) that since $w_{cc}(\eta) = 0$ is a trivial solution and since $D_{cc} = 0$ gives the buckling loads for the case when there are no in-span supports, the buckling loads are those values of $S_o = S_{o,n}$ that satisfy

$$S_{cc}(\eta_1, \eta_1) = 0 \tag{2.38}$$

Using the preceding reasoning for two in-span rigid supports, the buckling loads are the values of $S_o = S_{o,n}$ that satisfy

$$Z_{cc}(\eta_1, \eta_2) = 0 \tag{2.39}$$

Thus, buckling loads are those values that set the appropriate determinant of coefficients to zero.

2.3 Clamped-Free Beam

For the beam clamped at one end and free at the other end, we use Eq. (2.16) and the boundary conditions at $\eta = 1$ given in Table 2.1 for case 2. This yields the following system of equations:

$$\begin{aligned} w''(1) &= v'''(1,0)w''(0) + v''(1,0)w'''(0) + H''(1) \\ &\quad - K''(1) = 0 \\ w'''(1) + S_o w'(1) &= c_1 w''(0) + c_2 w'''(0) + H'''(1) + S_o H'(1) \\ &\quad - K'''(1) - S_o K'(1) = 0 \end{aligned} \tag{2.40}$$

where

$$\begin{aligned} c_1 &= S_o v''(1,0) + v''''(1,0) \\ c_2 &= S_o v'(1,0) + v'''(1,0) \end{aligned} \tag{2.41}$$

and

$$\begin{aligned} K'(1) &= K_1 w(\eta_1) v'(1,\eta_1) + K_2 w(\eta_2) v'(1,\eta_2) \\ K''(1) &= K_1 w(\eta_1) v''(1,\eta_1) + K_2 w(\eta_2) v''(1,\eta_2) \\ K'''(1) &= K_1 w(\eta_1) v'''(1,\eta_1) + K_2 w(\eta_2) v'''(1,\eta_2) \end{aligned} \tag{2.42}$$

Solving Eq. (2.40) for $w''(0)$ and $w'''(0)$, we obtain

$$\begin{aligned} w''(0) &= \frac{1}{D_{cf}}\Big(c_2\big(-H''(1)+K''(1)\big)+v''(1,0)\big(H'''(1) \\ &\quad + S_oH'(1) - K'''(1) - S_oK'(1)\big)\Big) \\ w'''(0) &= \frac{1}{D_{cf}}\Big(c_1\big(H''(1)-K''(1)\big)+v'''(1,0)\big(-H'''(1) \\ &\quad - S_oH'(1) + K'''(1) + S_oK'(1)\big)\Big) \end{aligned} \tag{2.43}$$

where

$$D_{cf} = c_2v'''(1,0) - c_1v''(1,0) \tag{2.44}$$

Substituting Eq. (2.43) into Eq. (2.16), using Eq. (2.42), and rearranging terms give

$$w_{cf}(\eta) = \frac{1}{D_{cf}}\Big(R_{cf}(\eta) + K_1w_{cf}(\eta_1)S_{cf}(\eta,\eta_1) + K_2w_{cf}(\eta_2)S_{cf}(\eta,\eta_2)\Big) \tag{2.45}$$

where

$$\begin{aligned} R_{cf} &= D_{cf}H(\eta) + P_{cf}(\eta) \\ P_{cf}(\eta) &= v'(\eta,0)\Big(-c_2H''(1)+v''(1,0)\big(H'''(1)+S_oH'(1)\big)\Big) \\ &\quad + v(\eta,0)\Big(c_1H''(1)-v'''(1,0)\big(H'''(1)+S_oH'(1)\big)\Big) \\ S_{cf}(\eta,\eta_j) &= -D_{cf}v(\eta,\eta_j)u(\eta-\eta_j) \\ &\quad + v'(\eta,0)\Big(c_2v''(1,\eta_j) - v''(1,0)\big(S_ov'(1,\eta_j)+v'''(1,\eta_j)\big)\Big) \\ &\quad + v(\eta,0)\Big(-c_1v''(1,\eta_j) + v'''(1,0)\big(S_ov'(1,\eta_j)+v'''(1,\eta_j)\big)\Big) \end{aligned} \tag{2.46}$$

To determine the solution for the case where there are no, one, or two in-span rigid supports, we notice that Eq. (2.45) has the same form as Eq. (2.22). Therefore, we replace the subscripts "*cc*" in Eqs. (2.24), (2.28), and (2.32), respectively, with "*cf*" to obtain the following displacement responses of the beam.

No in-span rigid supports

$$w_{cf}(\eta) = \frac{R_{cf}(\eta)}{D_{cf}} \tag{2.47}$$

One in-span rigid support

$$w_{cf}(\eta) = \frac{1}{D_{cf} S_{cf}(\eta_1, \eta_1)} \Big(R_{cf}(\eta) S_{cf}(\eta_1, \eta_1) - R_{cf}(\eta_1) S_{cf}(\eta, \eta_1) \Big) \tag{2.48}$$

Two in-span rigid supports

$$\begin{aligned} w_{cf}(\eta) = \frac{1}{D_{cf} E_{cf}(\eta_1, \eta_2)} \Big(& R_{cf}(\eta) E_{cf}(\eta_1, \eta_2) \\ & + E_{cf,1}(\eta_1, \eta_2) S_{cf}(\eta, \eta_1) + E_{cf,2}(\eta_1, \eta_2) S_{cf}(\eta, \eta_2) \Big) \end{aligned} \tag{2.49}$$

where

$$\begin{aligned} E_{cf}(\eta_1, \eta_2) &= S_{cf}(\eta_2, \eta_1) S_{cf}(\eta_1, \eta_2) - S_{cf}(\eta_1, \eta_1) S_{cf}(\eta_2, \eta_2) \\ &= D_{cf} Z_{cf}(\eta_1, \eta_2) \end{aligned} \tag{2.50}$$

and

$$\begin{aligned} Z_{cf}(\eta_1, \eta_2) = & \big(v(\eta_1, 0) v'(\eta_2, 0) - v'(\eta_1, 0) v(\eta_2, 0) \big) \times \\ & \Big(- v''(1, \eta_2) \big(v'''(1, \eta_1) + S_o v'(1, \eta_1) \big) \\ & + v''(1, \eta_1) \big(v'''(1, \eta_2) + S_o v'(1, \eta_2) \big) \Big) \\ & + v(\eta_2, \eta_1) \Big(v''(1, \eta_2) \big(c_1 v(\eta_1, 0) - c_2 v'(\eta_1, 0) \big) \\ & + \big(v''(1, 0) v'(\eta_1, 0) - v'''(1, 0) v(\eta_1, 0) \big) \times \\ & \big(v'''(1, \eta_2) + S_o v'(1, \eta_2) \big) \Big) \end{aligned} \tag{2.51}$$

2.3.1 Buckling Load

As noted previously, Eqs. (2.22) and (2.45) have the identical form, so that the subscript "*cc*" in Eq. (2.22) can be replaced with "*cf*" and the result is (2.45). Hence, the buckling loads are given for the respective cases by Eqs. (2.36), (2.38), and (2.39) with "*cc*" in these equations replaced by "*cf*." Thus, for no rigid in-span supports, the buckling equation is

$$D_{cf} = 0 \tag{2.52}$$

For example, if $K_f = 0$, then using Eqs. (2.41) and (2.44) and the functions appearing in the appropriate column of Table 2.3, we obtain

$$D_{cf} = \cos\sqrt{S_o} = 0 \tag{2.53}$$

Evaluation of Eq. (2.53) gives the lowest buckling load $S_{o,1} = 2.47$.

For one in-span rigid support located at η_1, the buckling equation is

$$S_{cf}(\eta_1, \eta_1) = 0 \tag{2.54}$$

and for two in-span rigid supports located at η_1 and η_2, the buckling equation is

$$Z_{cf}(\eta_1, \eta_2) = 0 \tag{2.55}$$

2.4 Clamped-Hinged Beam

For the beam clamped at one end and hinged at the other end, we use Eq. (2.16) and the boundary conditions at $\eta = 1$ given in Table 2.1 for case 3. This yields the following system of equations:

$$\begin{aligned} w(1) &= v'(1,0)w''(0) + v(1,0)w'''(0) + H(1) - K(1) = 0 \\ w''(1) &= v'''(1,0)w''(0) + v''(1,0)w'''(0) + H''(1) - K''(1) = 0 \end{aligned} \tag{2.56}$$

where $K''(1)$ is given by Eq. (2.42).

Solving Eq. (2.56) for $w''(0)$ and $w'''(0)$, we obtain

$$\begin{aligned} w''(0) &= \frac{1}{D_{ch}}\Big(v(1,0)\big(H''(1) - K''(1)\big) + v''(1,0)\big(-H(1) + K(1)\big)\Big) \\ w'''(0) &= \frac{1}{D_{ch}}\Big(v'(1,0)\big(-H''(1) + K''(1)\big) + v'''(1,0)\big(H(1) - K(1)\big)\Big) \end{aligned} \tag{2.57}$$

where

$$D_{ch} = v'(1,0)v''(1,0) - v(1,0)v'''(1,0) \tag{2.58}$$

Substituting Eq. (2.57) into Eq. (2.16), using the definitions of $K(1)$ and $K''(1)$, and rearranging terms give

$$w_{ch}(\eta) = \frac{1}{D_{ch}}\Big(R_{ch}(\eta) + K_1 w_{ch}(\eta_1)S_{ch}(\eta, \eta_1) + K_2 w_{ch}(\eta_2)S_{ch}(\eta, \eta_2)\Big) \tag{2.59}$$

where

$$
\begin{aligned}
R_{ch}(\eta) &= D_{ch}H(\eta) + P_{ch}(\eta) \\
P_{ch}(\eta) &= v'(\eta,0)\big(v(1,0)H''(1) - v''(1,0)H(1)\big) \\
&\quad - v(\eta,0)\big(v'(1,0)H''e(1) - v'''(1,0)H(1)\big) \\
S_{ch}(\eta,\eta_j) &= -D_{ch}v(\eta,\eta_j)u(\eta-\eta_j) \\
&\quad + v'(\eta,0)\big(v(1,\eta_j)v''(1,0) - v(1,0)v''(1,\eta_j)\big) \\
&\quad + v(\eta,0)\big(v'(1,0)v''(1,\eta_j) - v(1,\eta_j)v'''(1,0)\big)
\end{aligned}
\tag{2.60}
$$

To determine the solution for the case where there is either no, one, or two in-span rigid supports, we notice that Eq. (2.59) has the same form as Eq. (2.22). Therefore, we replace the subscripts "*cc*" in Eqs. (2.24), (2.28), and (2.32), respectively, with "*ch*" to obtain the following.

No in-span rigid supports

$$
w_{ch}(\eta) = \frac{R_{ch}(\eta)}{D_{ch}}
\tag{2.61}
$$

One in-span rigid support

$$
w_{ch}(\eta) = \frac{1}{D_{ch}S_{ch}(\eta_1,\eta_1)}\Big(R_{ch}(\eta)S_{ch}(\eta_1,\eta_1) - R_{ch}(\eta_1)S_{ch}(\eta,\eta_1)\Big)
\tag{2.62}
$$

Two in-span rigid supports

$$
\begin{aligned}
w_{ch}(\eta) &= \frac{1}{D_{ch}E_{ch}(\eta_1,\eta_2)}\Big(R_{ch}(\eta)E_{ch}(\eta_1,\eta_2) \\
&\quad + E_{ch,1}(\eta_1,\eta_2)S_{ch}(\eta,\eta_1) + E_{ch,2}(\eta_1,\eta_2)S_{ch}(\eta,\eta_2)\Big)
\end{aligned}
\tag{2.63}
$$

where

$$
\begin{aligned}
E_{ch}(\eta_1,\eta_2) &= S_{ch}(\eta_2,\eta_1)S_{ch}(\eta_1,\eta_2) - S_{ch}(\eta_1,\eta_1)S_{ch}(\eta_2,\eta_2) \\
&= D_{ch}Z_{ch}(\eta_1,\eta_2)
\end{aligned}
\tag{2.64}
$$

and

$$
\begin{aligned}
Z_{ch}(\eta_1,\eta_2) &= \big(-v(1,\eta_1)v''(1,\eta_2) + v''(1,\eta_1)v(1,\eta_2)\big)\times \\
&\quad \big(-v(\eta_1,0)v'(\eta_2,0) + v'(\eta_1,0)v(\eta_2,0)\big) \\
&\quad + v(\eta_2,\eta_1)\Big(v'(\eta_1,0)\big(v(1,0)v''(1,\eta_2) - v''(1,0)v(1,\eta_2)\big) \\
&\quad + v(\eta_1,0)\big(-v'(1,0)v''(1,\eta_2) + v'''e(1,0)v(1,\eta_2)\big)\Big)
\end{aligned}
\tag{2.65}
$$

2.4.1 Buckling Load

It is seen that Eqs. (2.22) and (2.59) have the identical form so that we can replace the subscript "*cc*" in Eq. (2.22) with "*ch*" and the result is Eq. (2.59). Hence, the buckling loads are given for the respective cases by Eqs. (2.36), (2.38), and (2.39) with "*cc*" in these equations replaced by "*ch*." Thus, for no rigid in-span supports, the buckling equation is

$$D_{ch} = 0 \tag{2.66}$$

For example, if $K_f = 0$, then using Eq. (2.58) and the functions appearing in the appropriate column of Table 2.3, we obtain

$$D_{ch} = \sin\sqrt{S_o} - \sqrt{S_o}\cos\sqrt{S_o} = 0 \tag{2.67}$$

Evaluation of Eq. (2.67) gives the lowest buckling load $S_{o,1} = 20.19$.

For one in-span rigid support located at η_1, the buckling equation is

$$S_{ch}(\eta_1, \eta_1) = 0 \tag{2.68}$$

and for two in-span rigid supports located at η_1 and η_2, the buckling equation is

$$Z_{ch}(\eta_1, \eta_2) = 0 \tag{2.69}$$

2.5 Hinged-Hinged Beam

For the beam hinged at both ends, we use Eq. (2.17) and the boundary conditions at $\eta = 1$ given in Table 2.1 for case 4. This yields the following system of equations:

$$\begin{aligned} w(1) &= d(1)w'(0) + v(1,0)w'''(0) + H(1) - K(1) = 0 \\ w''(1) &= d''(1)w'(0) + v''(1,0)w'''(0) + H''(1) - K''(1) = 0 \end{aligned} \tag{2.70}$$

where $K''(1)$ is given by Eq. (2.42) and $K(1)$ by Eq. (2.19).

Solving Eq. (2.70) for $w'(0)$ and $w'''(0)$, we obtain

$$\begin{aligned} w'(0) &= \frac{1}{D_{hh}}\Big(v(1,0)\big(H''(1) - K''(1)\big) + v''(1,0)\big(-H(1) + K(1)\big)\Big) \\ w'''(0) &= \frac{1}{D_{hh}}\Big(d(1)\big(-H''(1) + K''(1)\big) + d''(1)\big(H(1) - K(1)\big)\Big) \end{aligned} \tag{2.71}$$

where

$$D_{hh} = d(1)v''(1,0) - d''(1)v(1,0) \tag{2.72}$$

Substituting Eq. (2.71) into Eq. (2.17), using the definitions of $K(1)$ and $K''(1)$, and rearranging terms give

$$w_{hh}(\eta) = \frac{1}{D_{hh}}\Big(R_{hh}(\eta) + K_1 w_{hh}(\eta_1) S_{hh}(\eta,\eta_1) + K_2 w_{hh}(\eta_2) S_{hh}(\eta,\eta_2)\Big) \tag{2.73}$$

where

$$\begin{aligned} R_{hh}(\eta) &= D_{hh} H(\eta) + P_{hh}(\eta) \\ P_{hh}(\eta) &= v(\eta,0)\big(-d(1)H''(1) + d''(1)H(1)\big) \\ &\quad + d(\eta)\big(v(1,0)H''(1) - v''(1,0)H(1)\big) \\ S_{hh}(\eta,\eta_j) &= -D_{hh} v(\eta,\eta_j) u(\eta-\eta_j) \\ &\quad + v(\eta,0)\big(-d''(1)v(1,\eta_j) + d(1)v''(1,\eta_j)\big) \\ &\quad + d(\eta)\big(v(1,\eta_j)v''(1,0) - v(1,0)v''(1,\eta_j)\big) \end{aligned} \tag{2.74}$$

To determine the solution for the case where there is either no, one, or two in-span rigid supports, we notice that Eq. (2.73) has the same form as Eq. (2.22). Therefore, we can replace the subscripts "*cc*" in Eqs. (2.24), (2.28), and (2.32), respectively, with "*hh*" to obtain the following.

No in-span rigid supports

$$w_{hh}(\eta) = \frac{R_{hh}(\eta)}{D_{hh}} \tag{2.75}$$

One in-span rigid support

$$w_{hh}(\eta) = \frac{1}{D_{hh} S_{hh}(\eta_1,\eta_1)}\Big(R_{hh}(\eta) S_{hh}(\eta_1,\eta_1) - R_{hh}(\eta_1) S_{hh}(\eta,\eta_1)\Big) \tag{2.76}$$

Two in-span rigid supports

$$\begin{aligned} w_{hh}(\eta) = \frac{1}{D_{hh} E_{hh}(\eta_1,\eta_2)}\Big(& R_{hh}(\eta) E_{hh}(\eta_1,\eta_2) \\ & + E_{hh,1}(\eta_1,\eta_2) S_{hh}(\eta,\eta_1) + E_{hh,2}(\eta_1,\eta_2) S_{hh}(\eta,\eta_2)\Big) \end{aligned} \tag{2.77}$$

where

$$
\begin{aligned}
E_{hh}(\eta_1, \eta_2) &= S_{hh}(\eta_2, \eta_1) S_{hh}(\eta_1, \eta_2) - S_{hh}(\eta_1, \eta_1) S_{hh}(\eta_2, \eta_2) \\
&= D_{hh} Z_{hh}(\eta_1, \eta_2)
\end{aligned}
\tag{2.78}
$$

and

$$
\begin{aligned}
Z_{hh}(\eta_1, \eta_2) = &\big(-v(1, \eta_1)v''(1, \eta_2) + v''(1, \eta_1)v(1, \eta_2)\big)\times \\
&\big(-d(\eta_2)v(\eta_1, 0) + d(\eta_1)v(\eta_2, 0)\big) \\
&+ v(\eta_2, \eta_1)\Big(d(\eta_1)\big(v(1, 0)v''(1, \eta_2) - v''e(1, 0)v(1, \eta_2)\big) \\
&+ v(\eta_1, 0)\big(-d(1)v''(1, \eta_2) + d''(1)v(1, \eta_2)\big)\Big)
\end{aligned}
\tag{2.79}
$$

2.5.1 Buckling Load

It is seen that Eqs. (2.22) and (2.73) have the identical form so that we can replace the subscript "*cc*" in Eq. (2.22) with "*hh*" and the result is Eq. (2.73). Hence, the buckling loads are given for the respective cases by Eqs. (2.36), (2.38), and (2.39) with "*cc*" in these equations replaced by "*hh*." Thus, for no rigid in-span supports, the buckling equation is

$$
D_{hh} = 0 \tag{2.80}
$$

For example, if $K_f = 0$, then using Eqs. (2.72) and the functions appearing in the appropriate columns of Tables 2.3 and 2.4, we obtain

$$
D_{hh} = \sin\sqrt{S_o} = 0 \tag{2.81}
$$

Evaluation of Eq. (2.81) gives the lowest buckling load $S_{o,1} = 9.87$.

For one in-span rigid support located at η_1, the buckling equation is

$$
S_{hh}(\eta_1, \eta_1) = 0 \tag{2.82}
$$

and for two in-span rigid supports located at η_1 and η_2, the buckling equation is

$$
Z_{hh}(\eta_1, \eta_2) = 0 \tag{2.83}
$$

2.6 Numerical Results

We illustrate the methodology by determining the responses of the displacement, rotation, moment, and shear force of the beam as a function of position to various loads and boundary conditions. In addition, the lowest buckling load of the configuration is also determined and displayed in each figure. The results shown are all obtained from one interactive graphic, which can evaluate the response of the beam to all combinations of factors, boundary conditions, and applied forces and moments given in this chapter.

In all figures, the image on the top left shows the locations and relative magnitudes of the applied forces and the location of the rigid supports. The image on the top right gives the lowest buckling load for the selected configuration. The middle left image shows the nondimensional displacement $w(\eta)$ as a function of position, and the middle right image shows the nondimensional rotation $\Phi(\eta)$ as a function of position. The images on the lower left and the lower right, respectively, show the nondimensional moment $M(\eta)$ and the nondimensional shear force $V(\eta)$ as a function of position.

In Figs. 2.3, 2.4, 2.5, and 2.6, the same load is used to determine the response for four sets of boundary conditions when the beam has two in-span rigid supports. In Figs. 2.7, 2.8, and 2.9, we show the effects of the presence and absence of rigid supports on the response of a cantilever beam to the same applied loads. In Fig. 2.9, the second rigid support is placed at the free end of the beam, that is, $\eta_2 = 1$. This scenario produces the same results as those obtained for a beam clamped at one end and hinged at the other, including the buckling load. In all these figures, $S_o = 6$, which is a compressive axial force, and $K_f = 60$.

In Figs. 2.10 and 2.11, respectively, we show the effects that a compressive axial force and a tensile axial force of the same magnitude have on the response of the beam without in-span rigid supports for the same boundary conditions and loading.

From a qualitative examination of these figures, we find the following: (a) the boundary conditions are satisfied; (b) the displacements under the rigid supports are zero; (c) the displacement as a function of position is continuous; (d) the moment distribution shows an abrupt change in magnitude at those locations where an external moment has been applied; and (e) the shear force distribution shows an abrupt change in magnitude at those locations where a concentrated force has been applied and where there is a rigid support.

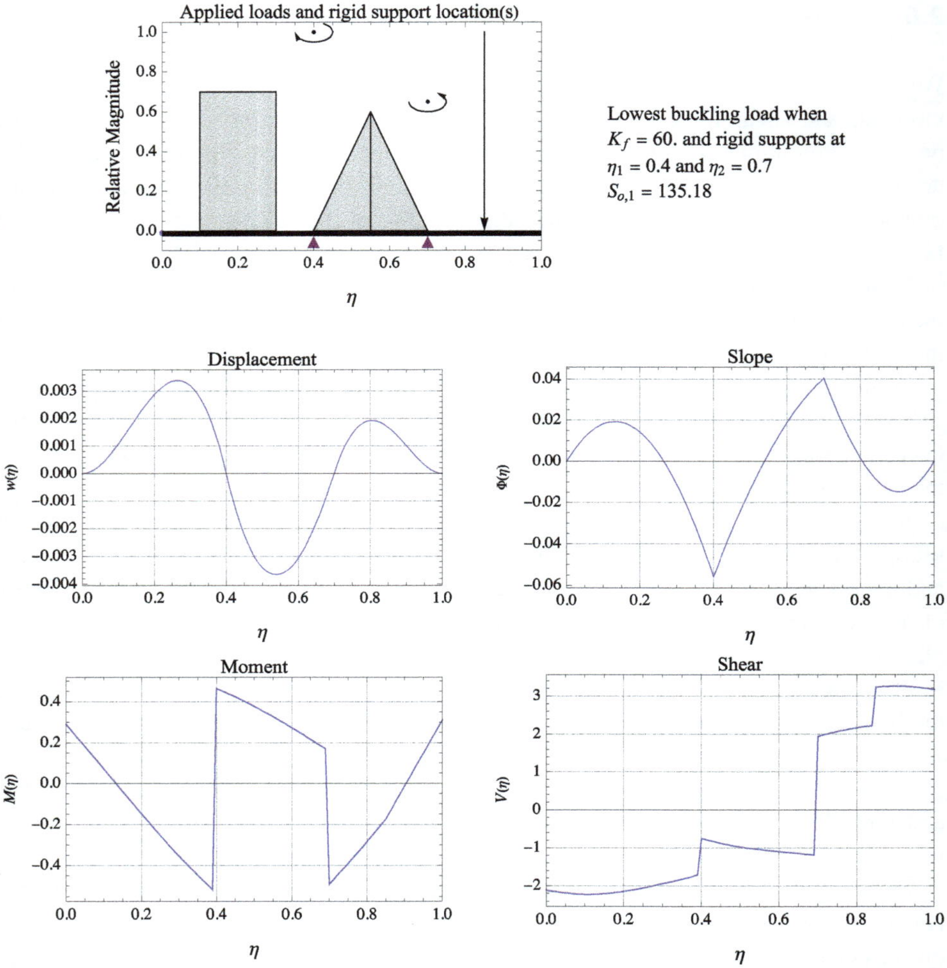

Fig. 2.3 Response of a uniform beam clamped at both ends with two in-span rigid supports. The beam is resting on an elastic foundation with $K_f = 60$, subjected to an axial compressive force $S_o = 6$, and subjected to the loads shown

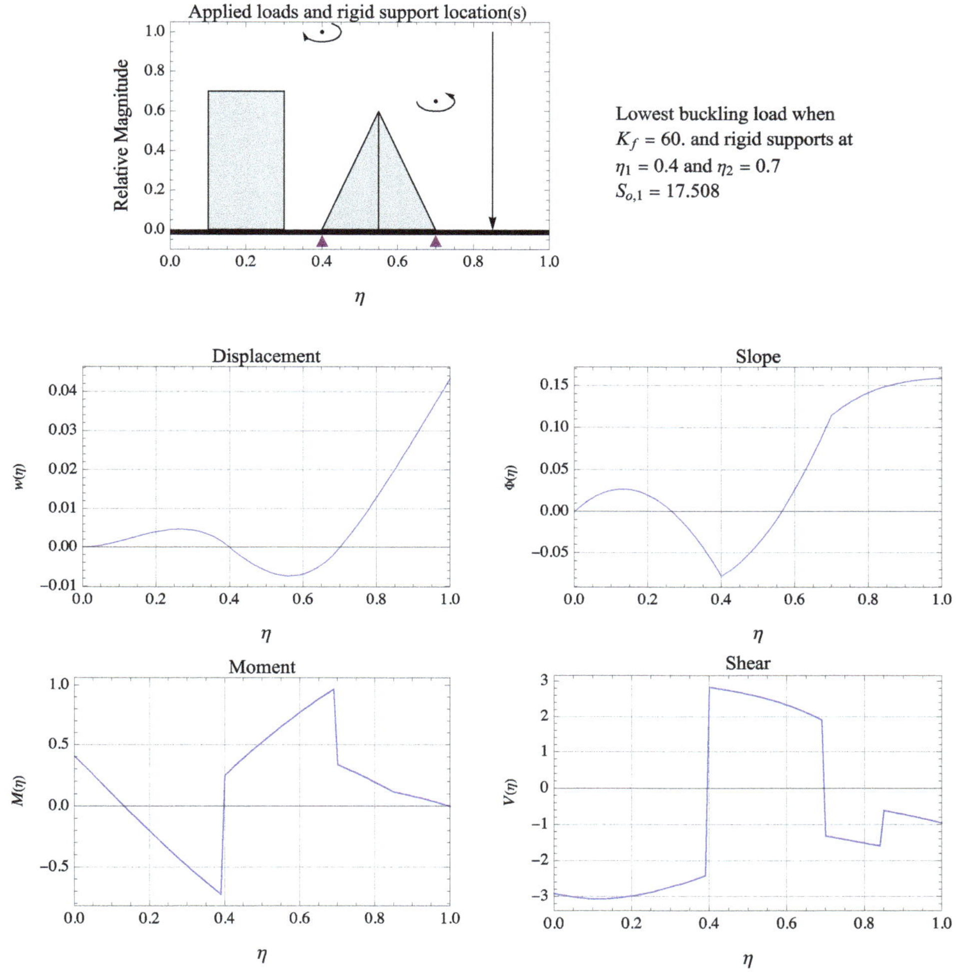

Fig. 2.4 Response of a uniform cantilever beam with two in-span rigid supports. The beam is resting on an elastic foundation with $K_f = 60$, subjected to an axial compressive force $S_o = 6$, and subjected to the applied loads shown. From Eqs. (2.9), (2.11), and Table 2.1, we see that $V(1) = -S_o\Phi(1) = -6 \times 0.158 = -0.948$, which confirms what is shown in the figure for $V(\eta)$

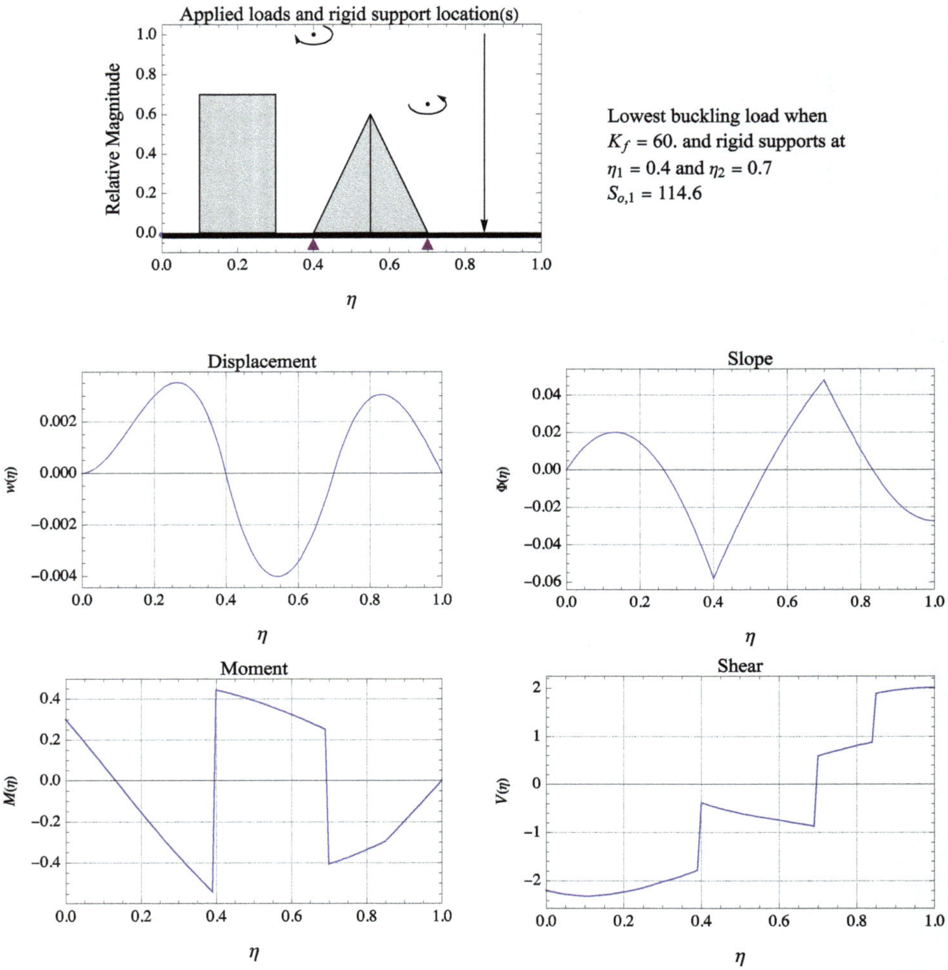

Fig. 2.5 Response of a uniform beam clamped at one end, hinged at the other end, and with two in-span rigid supports. The beam is resting on an elastic foundation with $K_f = 60$, subjected to an axial compressive force $S_o = 6$, and subjected to the applied loads shown

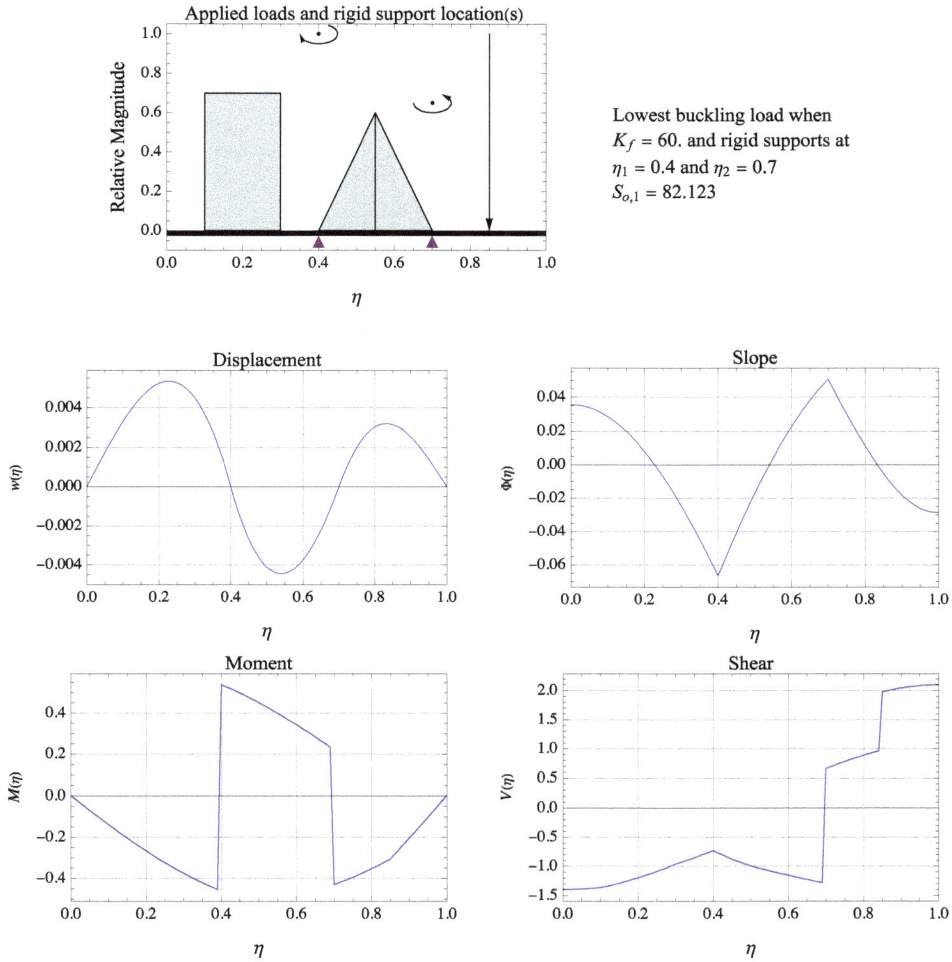

Fig. 2.6 Response of a uniform beam hinged at both ends with two in-span rigid supports. The beam is resting on an elastic foundation with $K_f = 60$, subjected to an axial compressive force $S_o = 6$, and subjected to the applied loads shown

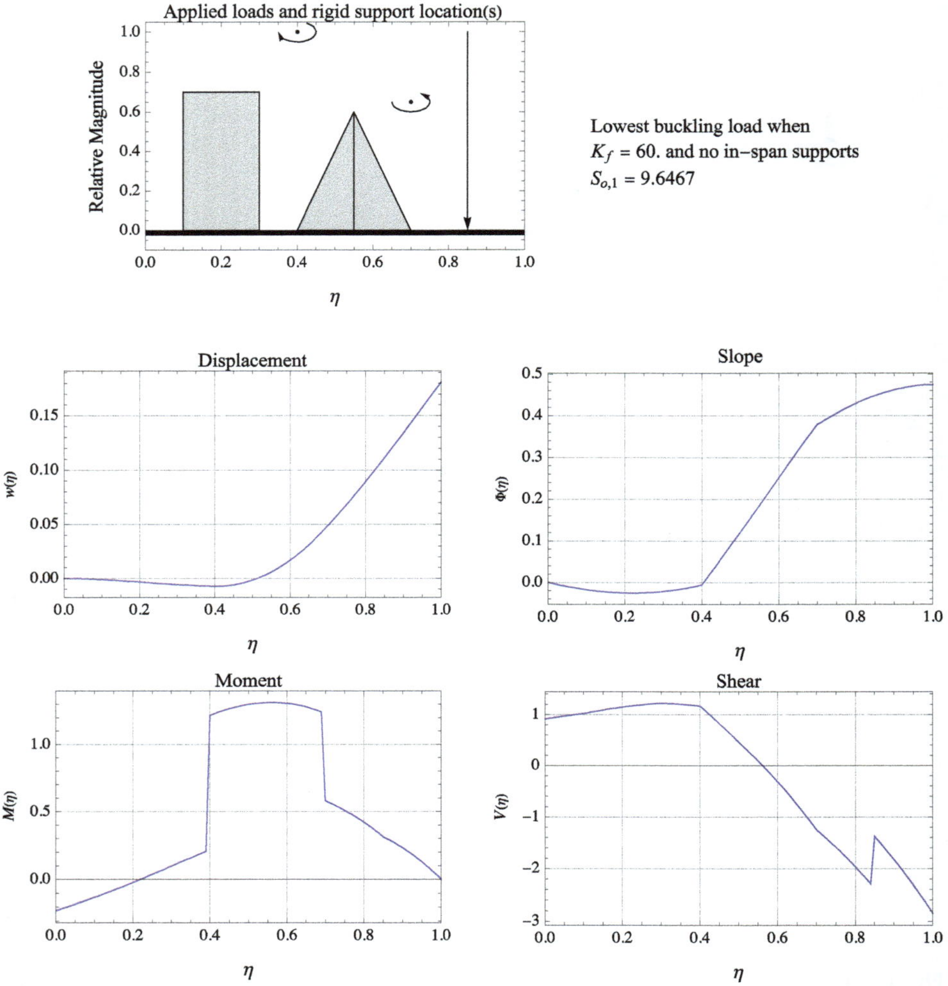

Fig. 2.7 Response of a uniform cantilever beam with no in-span rigid supports. The beam is resting on an elastic foundation with $K_f = 60$, subjected to an axial compressive force $S_o = 6$, and subjected to the applied loads shown. From Eqs. (2.9), (2.11), and Table 2.1, we see that $V(1) = -S_o\Phi(1) = -6 \times 0.474 = -2.84$, which confirms what is shown in the figure for $V(\eta)$

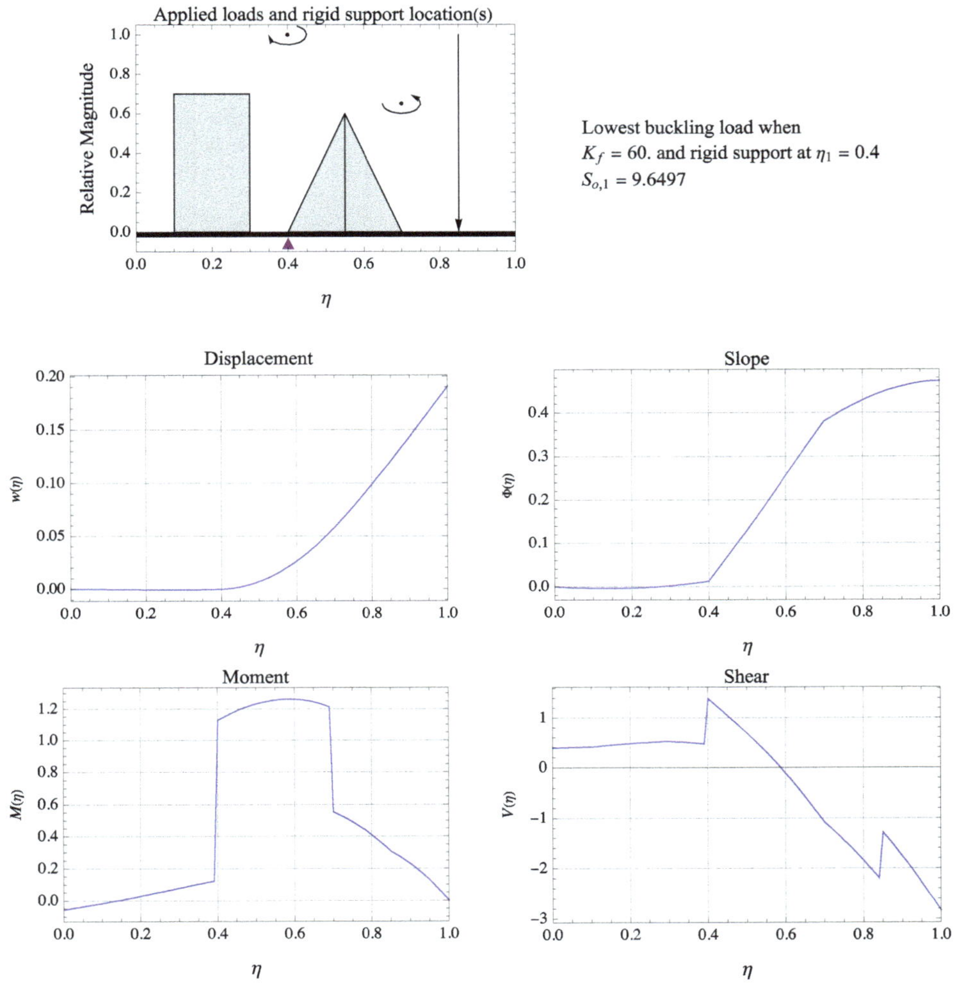

Fig. 2.8 Response of a uniform cantilever beam with one in-span rigid support. The beam is resting on an elastic foundation with $K_f = 60$, subjected to an axial compressive force $S_o = 6$, and subjected to the applied loads shown. From Eqs. (2.9), (2.11), and Table 2.1, we see that $V(1) = -S_o\Phi(1) = -6 \times 0.474 = -2.84$, which confirms what is shown in the figure for $V(\eta)$

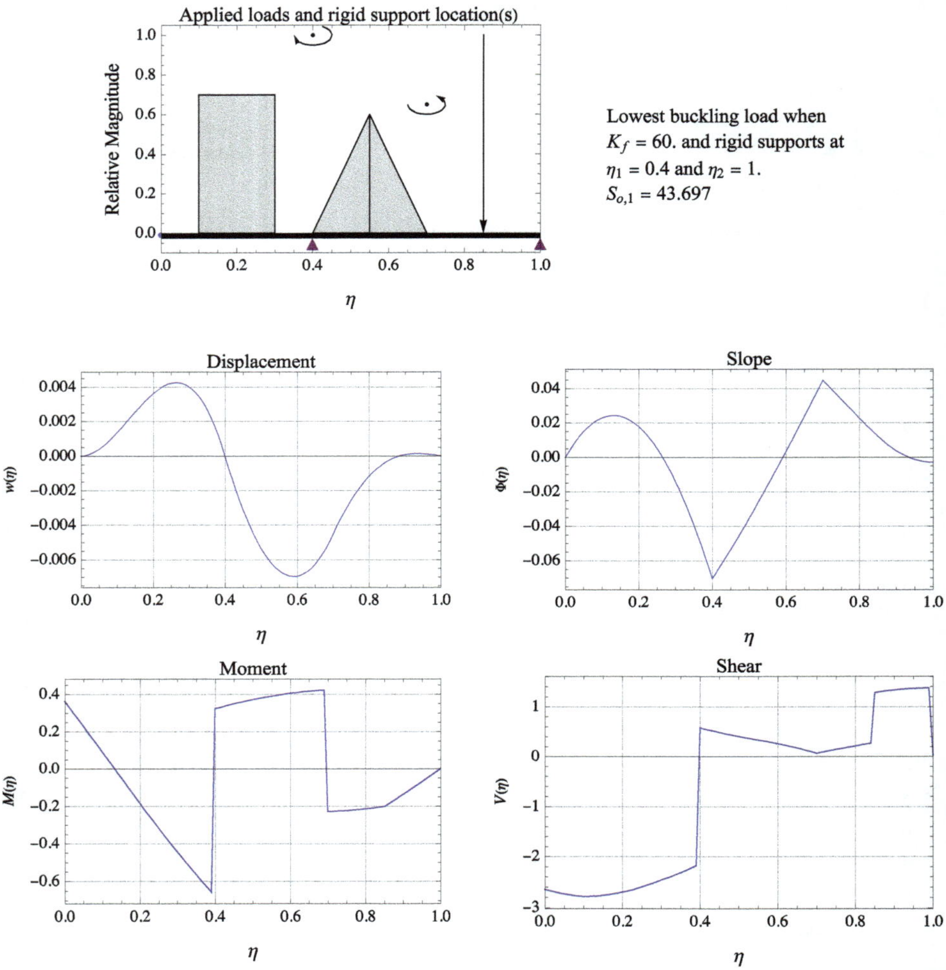

Fig. 2.9 Response of a uniform cantilever beam with two in-span rigid supports, one of which is placed at the free end of the beam. The beam is resting on an elastic foundation with $K_f = 60$, subjected to an axial compressive force $S_o = 6$, and subjected to the applied loads shown

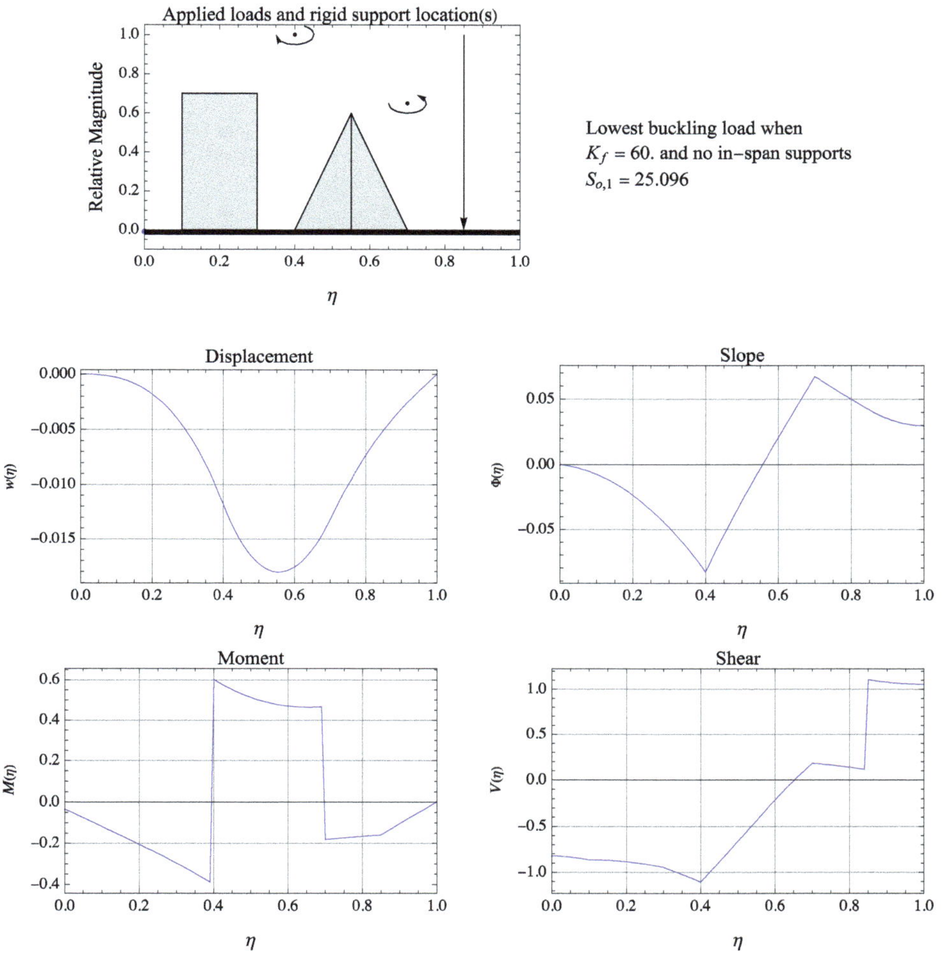

Fig. 2.10 Response of a uniform beam clamped at one end and hinged at the other end with no in-span rigid supports. The beam is resting on an elastic foundation with $K_f = 60$, subjected to an axial tensile force $S_o = -6$, and subjected to the applied loads shown. Compare with Fig. 2.11

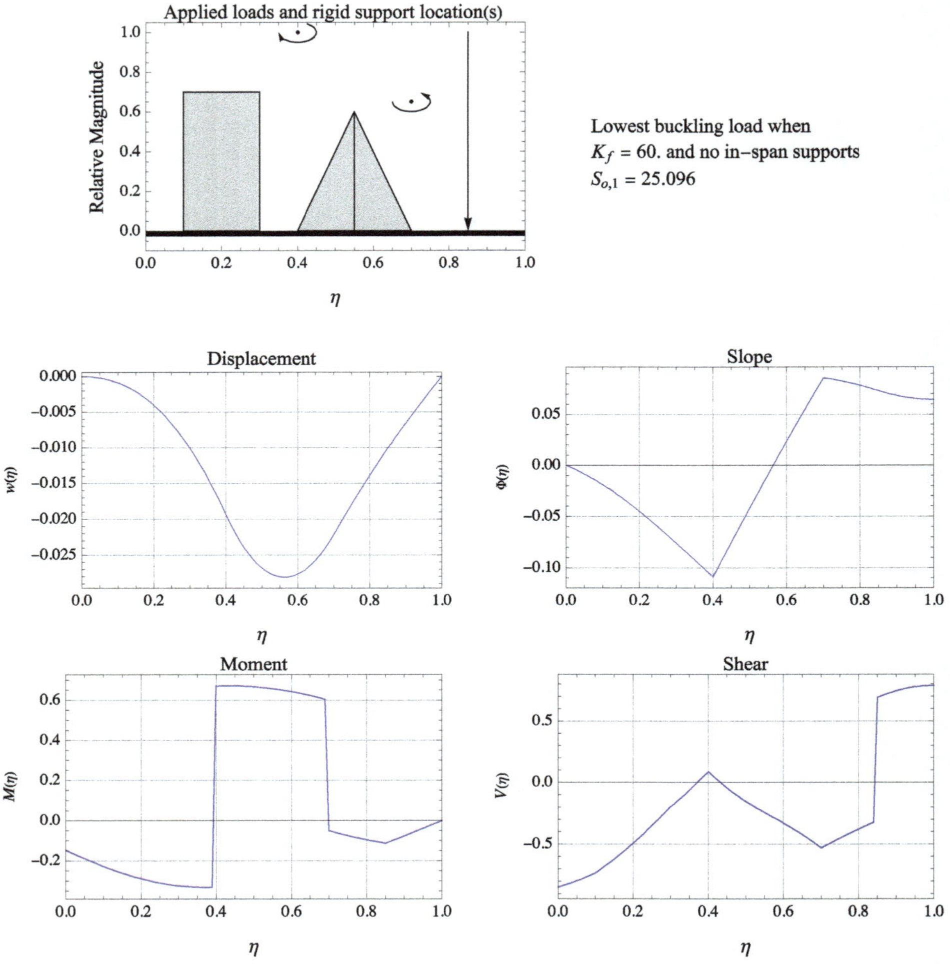

Fig. 2.11 Response of a uniform beam clamped at one end and hinged at the other end with no in-span rigid supports. The beam is resting on an elastic foundation with $K_f = 60$, subjected to an axial compressive force $S_o = 6$, and subjected to the applied loads shown. Compare with Fig. 2.10

References

1. Magrab EB (2024) Vibrations of elastic systems: with multiphysics applications, 2nd edn. Solid mechanics and its applications, vol 184. Springer, New York
2. Balachandran B, Magrab EB (2018) Vibrations, 3rd edn. Cambridge University, Cambridge

Beams with a Step Change in Flexural Rigidity

3

Abstract

Expressions for the spatial distribution of the displacement, rotation, moment, and shear force responses of a beam with a step change in flexural rigidity that is simultaneously subjected to a compressive axial force, concentrated forces, distributed forces, and the applied moments and rests on an elastic foundation and on in-span rigid supports are obtained for four sets of boundary conditions: clamped-clamped, clamped-free, clamped-hinged, and hinged-hinged. In addition, for any combination of these conditions, the lowest buckling load is determined. The solution to this beam configuration is obtained, in part, by using the solutions obtained in Chap. 2.

Keywords

Beams · Flexural rigidity · Axial force · Elastic foundation · Rigid supports · Concentrated force · Distributed force · Moments · Boundary conditions · Hinged · Clamped · Free · Buckling · Displacement · Shear force

3.1 Governing Equations

Consider a beam of length L (m) that is composed of two segments as shown in Fig. 3.1a. Segment 1 has a length L_{seg}, $0 < L_{seg} < L$, a tensile modulus E_1 (N/m^2), and moment of inertia I_1 (m^4). Its coordinate in the axial direction is denoted x_1, and it originates at the left end of the beam. The transverse displacement is $y_1(x_1)$ (m). The remaining

Supplementary Information The online version contains supplementary material available at (https://doi.org/10.1007/978-3-032-28726-7_3).

E. B. Magrab, *Static Analysis of Beams, Frames, and Arches*, Synthesis Lectures on Mechanical Engineering, https://doi.org/10.1007/978-3-032-28726-7_3

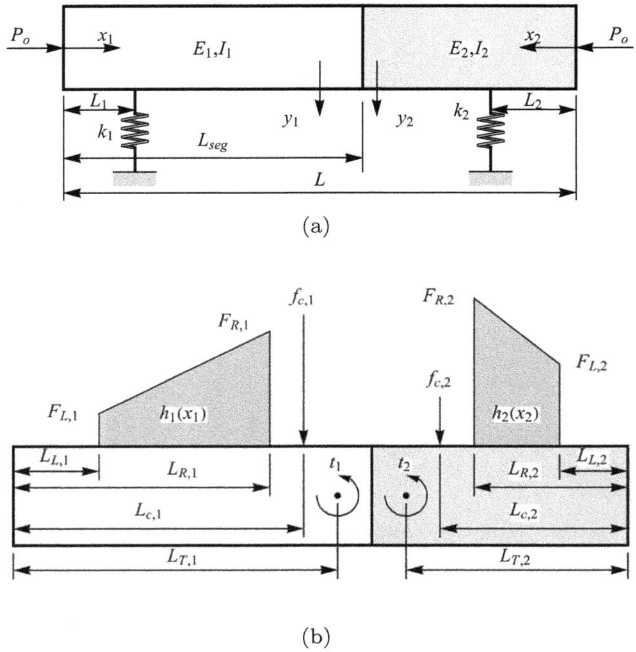

Fig. 3.1 (**a**) Nomenclature for a beam with step change in flexural rigidity with two in-span springs (elastic foundation not shown); (**b**) nomenclature for the trapezoidal distributed forces, concentrated forces, and counterclockwise moments

portion of the beam, segment 2, has a length $L - L_{seg}$, a tensile modulus E_2 (N/m^2), and a moment of inertia I_2 (m^4). Its coordinate in the axial direction is denoted x_2 and originates at the right end of the beam. The transverse displacement is $y_2(x_2)$ (m). The beam is subjected to a compressive axial force P_o (N) and rests on an elastic foundation of stiffness k_f (N/m^2). In addition, segment 1 has a linear elastic spring k_1 (N/m) attached at $x_1 = L_1, 0 \le L_1 \le L_{seg}$, and segment 2 has a linear elastic spring k_2 (N/m) attached at $x_2 = L_2, 0 \le L_2 \le L - L_{seg}$. The introduction of these in-span springs will be used at an appropriate point in the solution process to obtain two rigid in-span supports, one in each segment, by letting $k_j \to \infty$.

Referring to Fig. 3.1b, the beam is subject to the following applied loads. Segment 1 is subjected to a concentrated load $f_{c,1}$ at $x_1 = L_{c,1}, 0 < L_{c,1} \le L_{seg}$, a trapezoidal load $h_1(x_1)$ (N/m), where $0 \le x_1 \le L_{seg}$, and a counterclockwise moment t_1 (Nm) at $x = L_{T,1}$ (m), $0 \le L_{T,1} \le L_{seg}$. Similarly, segment 2 is subjected to a concentrated load $f_{c,2}$ at $x_2 = L_{c,2}, 0 \le L_{c,2} \le L - L_{seg}$, a trapezoidal load $h_2(x_2)$ (N/m) where $0 \le x_2 \le L - L_{seg}$, and a counterclockwise moment t_2 (Nm) at $x = L_{T,2}$ (m), $0 \le L_{T,2} \le L - L_{seg}$.

The governing equation for beam segment 1, which is valid for $0 \le x_1 \le L_{seg}$, can be written as

$$\begin{aligned} E_1 I_1 \frac{d^4 y_1}{dx_1^4} + P_o \frac{d^2 y_1}{dx_1^2} + k_f y_1 + k_1 \delta(x_1 - L_1) y_1 = h_1(x_1) \\ + f_{c,1} \delta(x_1 - L_{c,1}) - t_1 \frac{d}{dx_1} \delta(x_1 - L_{T,1}) \quad \text{N/m} \end{aligned} \tag{3.1}$$

and that for beam segment 2, which is valid for $0 \le x_2 \le L - L_{seg}$, can be written as

$$\begin{aligned} E_2 I_2 \frac{d^4 y_2}{dx_2^4} + P_o \frac{d^2 y_2}{dx_2^2} + k_f y_2 + k_2 \delta(x_2 - L_2) y_2 \\ = h_2(x_2) + f_{c,2} \delta(x_2 - L_{c,2}) - t_2 \frac{d}{dx_2} \delta(x_2 - L_{T,2}) \quad \text{N/m} \end{aligned} \tag{3.2}$$

where

$$h_j(x_j) = (a_j x_j + b_j)\big(u(x_j - L_{L,j}) - u(x_j - L_{R,j})\big) \quad j = 1, 2 \tag{3.3}$$

and

$$\begin{aligned} a_j &= \frac{F_{L,j} - F_{R,j}}{L_{L,j} - L_{R,j}} \quad \text{N/m}^2 \\ b_j &= \frac{F_{R,j} L_{L,j} - F_{L,j} L_{R,j}}{L_{L,j} - L_{R,j}} \quad \text{N/m} \end{aligned} \tag{3.4}$$

In Eq. (3.4), $F_{L,j}$ (N/m) and $F_{R,j}$ (N/m), respectively, are the magnitudes of the end points of the trapezoidal load located at $L_{L,j}$ and $L_{R,j}$, where $L_{L,j} < L_{R,j}$. In addition, the slope of each beam segment is $\phi_j = dy_j/dx_j$, $j = 1, 2$, the moment is $m_j = E_j I_j d^2 y_j/dx_j^2$ (Nm), and the shear force is $v_{s,j} = E_j I_j d^3 y_j/dx_j^3$ (N).

To place Eqs. (3.1)–(3.4) in a more useful form, we introduce the following nondimensional quantities $\eta = x_1/L$, $\xi = x_2/L$, and

$$w_j = \frac{E_1 I_1 y_j}{F_{max} L^3} \quad j = 1, 2 \tag{3.5}$$

where F_{max} (N) will be defined subsequently. Upon substituting these definitions into Eqs. (3.1) and (3.2), respectively, we arrive at

$$\begin{aligned} \frac{d^4 w_1}{d\eta^4} + S_o \frac{d^2 w_1}{d\eta^2} + K_f w_1 + K_1 \delta(\eta - \eta_1) w_1 = g_1(\eta) \\ + F_{c,1} \delta(\eta - \eta_{c,1}) - T_1 \frac{d}{d\eta} \delta(\eta - \eta_{T,1}) \quad 0 \le \eta \le a_s \end{aligned} \tag{3.6}$$

and

$$\begin{aligned}\frac{d^4 w_2}{d\xi^4} + \beta S_o \frac{d^2 w_2}{d\xi^2} + \beta K_f w_2 + \beta K_2 \delta(\xi - \xi_2) w_2 = \beta g_2(\xi) \\ + \beta F_{c,2} \delta(\xi - \xi_{c,2}) - \beta T_2 \frac{d}{d\xi} \delta(\xi - \xi_{T,2}) \quad 0 \le \xi \le b_s\end{aligned} \tag{3.7}$$

where $\beta = E_1 I_1 / E_2 I_2$, $a_s = L_{seg}/L$, $b_s = 1 - a_s$, $\eta_1 = L_1/L$, $\xi_2 = L_2/L$, $\eta_{c,1} = L_{c,1}/L$, $\xi_{c,2} = L_{c,2}/L$, $\eta_{T,1} = L_{T,1}/L$, $\xi_{T,2} = L_{T,2}/L$, and

$$\begin{aligned} S_o &= \frac{P_o L^2}{E_1 I_1} \qquad K_j = \frac{k_j L^3}{E_1 I_1} \qquad K_f = \frac{k_f L^4}{E_1 I_1} \\ F_{c,j} &= \frac{f_{c,j}}{F_{max}} \qquad T_j = \frac{t_j}{F_{max} L} \end{aligned} \tag{3.8}$$

are nondimensional quantities. In addition, Eqs. (3.3) and (3.4) for $j = 1$ become

$$g_1(\eta) = \frac{L}{F_{max}} h_1(x_1) = (\tilde{a}_1 \eta + \tilde{b}_1)\big(u(\eta - \eta_{L,1}) - u(\eta - \eta_{R,1})\big) \tag{3.9}$$

where

$$\begin{aligned} \tilde{a}_1 &= \frac{q_{L,1} - q_{R,1}}{\eta_{L,1} - \eta_{R,1}} \\ \tilde{b}_1 &= \frac{q_{R,1}\eta_{L,j} - q_{L,1}\eta_{R,1}}{\eta_{L,1} - \eta_{R,1}} \end{aligned} \tag{3.10}$$

and $\eta_{L,1} = L_{L,1}/L$, $\eta_{R,1} = L_{R,1}/L$, $q_{L,1} = LF_{L,1}/F_{max}$, and $q_{R,1} = LF_{R,1}/F_{max}$. For $j = 2$, Eqs. (3.3) and (3.4) become

$$g_2(\xi) = \frac{L}{F_{max}} h_2(x_2) = (\tilde{a}_2 \xi + \tilde{b}_2)\big(u(\xi - \xi_{L,1}) - u(\xi - \xi_{R,1})\big) \tag{3.11}$$

where

$$\begin{aligned} \tilde{a}_2 &= \frac{q_{L,2} - q_{R,2}}{\xi_{L,2} - \xi_{R,2}} \\ \tilde{b}_2 &= \frac{q_{R,2}\xi_{L,2} - q_{L,2}\xi_{R,2}}{\xi_{L,2} - \xi_{R,2}} \end{aligned} \tag{3.12}$$

and $\xi_{L,2} = L_{L,2}/L$, $\xi_{R,2} = L_{R,2}/L$, $q_{L,2} = LF_{L,2}/F_{max}$, and $q_{R,2} = LF_{R,2}/F_{max}$.

It is mentioned that the applied loads cannot be made independent of the beam segments, since each segment is defined by its own independent coordinate system.

It is noted that when $\beta < 1$, segment 2 is stiffer than segment 1, and when $\beta > 1$, segment 1 is stiffer than segment 2. When $\beta = 1$, we have the case of a uniform beam, which has been discussed in Chap. 2.

The quantity F_{max} is selected as the maximum value of the following quantities: $F_{c,j}$ (N), $LF_{L,j}$ (N), t_j/L (N), and $LF_{R,j}$ (N), $j = 1, 2$. Therefore, depending on which entities are selected, at least one of the quantities will have the value of 1, and the rest of them will have a value less than or equal to 1. Thus, when more than one of these loads exists simultaneously, one must have an estimate of their respective magnitudes and the length of the beam.

In terms of nondimensional quantities, we find that the slope of beam segment 1, denoted ϕ_1, becomes

$$\Phi_1 = \frac{E_1 I_1 \phi_1}{F_{max} L^2} = \frac{dw_1}{d\eta} \tag{3.13}$$

The moment m_1 (Nm) becomes

$$M_1 = \frac{m_1}{F_{max} L} = \frac{d^2 w_1}{d\eta^2} \tag{3.14}$$

and the shear force $v_{s,1}$ (N) becomes

$$V_1 = \frac{v_{s,1}}{F_{max}} = \frac{d^3 w_1}{d\eta^3} \tag{3.15}$$

The slope of the beam segment 2, denoted ϕ_2, becomes

$$\Phi_2 = \frac{E_1 I_1 \phi_2}{F_{max} L^2} = \frac{dw_2}{d\xi} \tag{3.16}$$

the moment m_2 (Nm) becomes

$$M_2 = \frac{m_2}{F_{max} L} = \frac{1}{\beta} \frac{d^2 w_2}{d\xi^2} \tag{3.17}$$

and the shear force $v_{s,2}$ (N) becomes

$$V_2 = \frac{v_{s,2}}{F_{max}} = \frac{1}{\beta} \frac{d^3 w_2}{d\xi^3} \tag{3.18}$$

The quantities given by Eqs. (3.13)–(3.18) are the quantities for which numerical values are determined. These relations are also used to convert the numerical values into dimensional quantities.

The boundary conditions that will be considered are given in Table 3.1.

Table 3.1 Boundary conditions for a beam with a step change in flexural rigidity

Case	Boundary conditions	$\eta = 0\ [w_1(\eta)]$	$\xi = 0\ [w_2(\xi)]$
1	Clamped-clamped	$w_1(0) = 0$ $w_1'(0) = 0$	$w_2(0) = 0$ $w_2'(0) = 0$
2	Clamped-free	$w_1(0) = 0$ $w_1'(0) = 0$	$w_2''(0) = 0$ $w_2'''(0) + \beta S_o w_2'(0) = 0$
3	Clamped-hinged	$w_1(0) = 0$ $w_1'(0) = 0$	$w_2(0) = 0$ $w_2''(0) = 0$
4	Hinged-hinged	$w_1(0) = 0$ $w_1''(0) = 0$	$w_2(0) = 0$ $w_2''(0) = 0$

In addition to the boundary conditions, there are continuity conditions that must be satisfied at the interface of the two segments. These continuity conditions are

$$\begin{aligned} y_1(L_{seg}) &= y_2(L - L_{seg}) \\ \phi_1(L_{seg}) &= -\phi_2(L - L_{seg}) \\ m_1(L_{seg}) &= m_2(L - L_{seg}) \\ v_{s,1}(L_{seg}) &= -v_{s,2}(L - L_{seg}), \end{aligned} \tag{3.19}$$

which, in terms of the nondimensional quantities, can be written as

$$\begin{aligned} w_1|_{\eta=a_s} &= w_2|_{\xi=b_s} \\ \left.\frac{dw_1}{d\eta}\right|_{\eta=a_s} &= -\left.\frac{dw_2}{d\xi}\right|_{\xi=b_s} \\ \beta\left.\frac{d^2 w_1}{d\eta^2}\right|_{\eta=a_s} &= \left.\frac{d^2 w_2}{d\xi^2}\right|_{\xi=b_s} \\ \beta\left.\frac{d^3 2w_1}{d\eta^3}\right|_{\eta=a_s} &= -\left.\frac{d^3 w_2}{d\xi^3}\right|_{\xi=b_s} \end{aligned} \tag{3.20}$$

Upon comparing Eqs. (3.6) and (3.7) with Eq. (2.5), we can use Eq. (2.12) to obtain the solutions to Eqs. (3.6) and (3.7), respectively, as

$$\begin{aligned} w_1(\eta) &= d'(\eta)w_1(0) + d(\eta)w_1'(0) + v'(\eta,0)w_1''(0) \\ &\quad + v(\eta,0)w_1'''(0) + H_1(\eta) - K_{s,1}(\eta) \\ w_2(\xi) &= \hat{d}'(\xi)w_2(0) + \hat{d}(\xi)w_2'(0) + \hat{v}'(\xi,0)w_2''(0) \\ &\quad + \hat{v}(\xi,0)w_2'''(0) + \beta\hat{H}_2(\xi) - \beta\hat{K}_{s,2}(\xi) \end{aligned} \tag{3.21}$$

where

$$
\begin{aligned}
H_1(\eta) &= \tilde{a}_1 A_1(\eta) + \tilde{b}_1 B_1(\eta) + F_{c,1} v(\eta, \eta_{c,1}) \theta(\eta - \eta_{c,1}) \\
&\quad - T_1 v'(\eta, \eta_{T,1}) \theta(\eta - \eta_{T,1}) \\
\hat{H}_2(\xi) &= \tilde{a}_2 \hat{A}_2(\xi) + \tilde{b}_2 \hat{B}_2(\xi) + F_{c,2} \hat{v}(\xi, \xi_{c,2}) \theta(\xi - \xi_{c,2}) \\
&\quad - T_2 \hat{v}'(\xi, \xi_{T,2}) \theta(\xi - \xi_{T,2}) \\
K_{s,1}(\eta) &= K_1 w_1(\eta_1) v(\eta, \eta_1) \theta(\eta - \eta_1) \\
\hat{K}_{s,2}(\xi) &= K_2 w_2(\xi_2) \hat{v}(\xi, \xi_2) \theta(\xi - \xi_2)
\end{aligned} \tag{3.22}
$$

and

$$
\begin{aligned}
A_1(\eta) &= \big(\eta_{L,1} r'(\eta, \eta_{L,1}) + r(\eta, \eta_{L,1})\big) \theta(\eta - \eta_{L,1}) \\
&\quad - \big(\eta_{R,1} r'(\eta, \eta_{R,1}) + r(\eta, \eta_{R,1})\big) \theta(\eta - \eta_{R,1}) \\
B_1(\eta) &= r'(\eta, \eta_{L,1}) \theta(\eta - \eta_{L,1}) - r'(\eta, \eta_{R,1}) \theta(\eta - \eta_{R,1}) \\
\hat{A}_2(\xi) &= \big(\xi_{L,2} \hat{r}'(\xi, \xi_{L,2}) + \hat{r}(\xi, \xi_{L,2})\big) \theta(\xi - \xi_{L,2}) \\
&\quad - \big(\xi_{R,2} \hat{r}'(\xi, \xi_{R,2}) + \hat{r}(\xi, \xi_{R,2})\big) \theta(\xi - \xi_{R,2}) \\
\hat{B}_2(\xi) &= \hat{r}'(\xi, \xi_{L,2}) \theta(\xi - \xi_{L,2}) - \hat{r}'(\xi, \xi_{R,2}) \theta(\xi - \xi_{R,2})
\end{aligned} \tag{3.23}
$$

In Eq. (3.21), the solution functions $r(\xi, \alpha))$,$v(\xi, \alpha)$, and $d(\xi)$ and their derivatives, respectively, are given in Tables 2.2, 2.3, and 2.4 and are a shorthand notation as indicated in Eq. (2.15). The solution functions $\hat{r}(\xi, \alpha))$,$\hat{v}(\xi, \alpha)$, and $\hat{d}(\xi)$ and their derivatives are a shorthand notation for

$$
\begin{aligned}
\hat{r}(\xi, \alpha) &= r(\xi, \alpha, \beta S_o, \beta K_f) \\
\hat{v}(\xi, \alpha) &= v(\xi, \alpha, \beta S_o, \beta K_f) \\
\hat{d}(\xi) &= d(\xi, \beta S_o, \beta K_f)
\end{aligned} \tag{3.24}
$$

and are given in Tables 2.2, 2.3, and 2.4 with η replaced by ξ, S_o replaced by βS_o, and K_f replaced by βK_f. In other words, the quantities without the hat (^) indicate functions of η and those with the hat indicate functions of ξ.

We now use the boundary conditions given in Table 3.1 to simplify Eq. (3.21). The results are given in Table 3.2. We see from the expressions in Table 3.2 that each pair has the form

$$
\begin{aligned}
w_1(\eta) &= c_{11} p_{11}(\eta) + c_{12} p_{12}(\eta) + H_1(\eta) - K_{s,1}(\eta) \\
w_2(\xi) &= c_{21} \hat{p}_{21}(\xi) + c_{22} \hat{p}_{22}(\xi) + \beta \hat{H}_2(\xi) - \beta \hat{K}_{s,2}(\xi)
\end{aligned} \tag{3.25}
$$

Table 3.2 Equation (3.21) simplified for the boundary conditions given in Table 3.1

Case	Boundary conditions	Simplified Eq. (3.21)
1	Clamped-clamped	$w_1(\eta) = v'(\eta, 0)w_1''(0) + v(\eta, 0)w_1'''(0) + H_1(\eta) - K_{s,1}(\eta)$ $w_2(\xi) = \hat{v}'(\xi, 0)w_2''(0) + \hat{v}(\xi, 0)w_2'''(0) + \beta\hat{H}_2(\xi) - \beta\hat{K}_{s,2}(\xi)$
2	Clamped-free	$w_1(\eta) = v'(\eta, 0)w_1''(0) + v(\eta, 0)w_1'''(0) + H_1(\eta) - K_{s,1}(\eta)$ $w_2(\xi) = \hat{d}'(\xi)w_2(0) + \big(\hat{d}(\xi) - \beta S_o\hat{v}(\xi, 0)\big)w_2'(0)) + \beta\hat{H}_2(\xi)$ $-\beta\hat{K}_{s,2}(\xi)$
3	Clamped-hinged	$w_1(\eta) = v'(\eta, 0)w_1''(0) + v(\eta, 0)w_1'''(0) + H_1(\eta) - K_{s,1}(\eta)$ $w_2(\xi) = \hat{d}(\xi)w_2'(0) + \hat{v}(\xi, 0)w_2'''(0) + \beta\hat{H}_2(\xi) - \beta\hat{K}_{s,2}(\xi)$
4	Hinged-hinged	$w_1(\eta) = d(\eta)w_1'(0) + v(\eta, 0)w_1'''(0) + H_1(\eta) - K_{s,1}(\eta)$ $w_2(\xi) = \hat{d}(\xi)w_2'(0) + \hat{v}(\xi, 0)w_2'''(0) + \beta\hat{H}_2(\xi) - \beta\hat{K}_{s,2}(\xi)$

Table 3.3 Definitions of the terms in Eq. (3.25) for the boundary conditions given in Table 3.1

Case	Boundary conditions	c_{ij}	p_{ij}
1	Clamped-clamped	$c_{11} = w_1''(0)$ $c_{12} = w_1'''(0)$ $c_{21} = w_2''(0)$ $c_{22} = w_2'''(0)$	$p_{11}(\eta) = v'(\eta, 0)$ $p_{12}(\eta) = v(\eta, 0)$ $p_{21}(\xi) = \hat{v}'(\xi, 0)$ $\hat{p}_{22}(\xi) = \hat{v}(\xi, 0)$
2	Clamped-free	$c_{11} = w_1''(0)$ $c_{12} = w_1'''(0)$ $c_{21} = w_2(0)$ $c_{22} = w_2'(0)$	$p_{11}(\eta) = v'(\eta, 0)$ $p_{12}(\eta) = v(\eta, 0)$ $p_{21}(\xi) = \hat{d}'(\xi)$ $\hat{p}_{22}(\xi) = \hat{d}(\xi) - \beta S_o\hat{v}(\xi, 0)$
3	Clamped-hinged	$c_{11} = w_1''(0)$ $c_{12} = w_1'''(0)$ $c_{21} = w_2'(0)$ $c_{22} = w_2'''(0)$	$p_{11}(\eta) = v'(\eta, 0)$ $p_{12}(\eta) = v(\eta, 0)$ $p_{21}(\xi) = \hat{d}(\xi)$ $\hat{p}_{22}(\xi) = \hat{v}(\xi, 0)$
4	Hinged-hinged	$c_{11} = w_1'(0)$ $c_{12} = w_1'''(0)$ $c_{21} = w_2'(0)$ $c_{22} = w_2'''(0)$	$p_{11}(\eta) = d(\eta)$ $p_{12}(\eta) = v(\eta, 0)$ $p_{21}(\xi) = \hat{d}(\xi)$ $\hat{p}_{22}(\xi) = \hat{v}(\xi, 0)$

where the definitions of c_{ij} and p_{ij} are given in Table 3.3. To determine the unknown constants c_{ij}, we use Eq. (3.25) in the continuity conditions given by Eq. (3.20). Upon substituting Eq. (3.25) in Eq. (3.20), we obtain

$$\mathbf{Ac} = \mathbf{b} \tag{3.26}$$

where

$$\mathbf{A} = \begin{pmatrix} p_{11}(a_s) & p_{12}(a_s) & -\hat{p}_{21}(b_s) & -\hat{p}_{22}(b_s) \\ p'_{11}(a_s) & p'_{12}(a_s) & \hat{p}'_{21}(b_s) & \hat{p}'_{22}(b_s) \\ \beta p''_{11}(a_s) & \beta p''_{12}(a_s) & -\hat{p}''_{21}(b_s) & -\hat{p}''_{22}(b_s) \\ \beta p'''_{11}(a_s) & \beta p'''_{12}(a_s) & \hat{p}'''_{21}(b_s) & \hat{p}'''_{22}(b_s) \end{pmatrix} \qquad \mathbf{c} = \begin{Bmatrix} c_{11} \\ c_{12} \\ c_{21} \\ c_{22} \end{Bmatrix} \tag{3.27}$$

and

$$\mathbf{b} = \begin{Bmatrix} -H_1(a_s) + \beta \hat{H}_2(b_s) + K_{s,1}(a_s) - \beta \hat{K}_{s,2}(b_s) \\ -H'_1(a_s) - \beta \hat{H}'_2(b_s) + K'_{s,1}(a_s) + \beta \hat{K}'_{s,2}(b_s) \\ -\beta H''_1(a_s) + \beta \hat{H}''_2(b_s) + \beta K''_{s,1}(a_s) - \beta \hat{K}''_{s,2}(b_s) \\ -\beta H'''_1(a_s) - \beta \hat{H}'''_2(b_s) + \beta K'''_{s,1}(a_s) + \beta \hat{K}'''_{s,2}(b_s) \end{Bmatrix} \tag{3.28}$$

where the prime ($'$) represents the derivative with respect to its original argument, either η or ξ. The elements of **b** are b_j, $j = 1, 2, 3, 4$.

Getting an algebraic solution to Eq. (3.26) is not practical. Instead, we can get several very useful results by employing a hybrid approach that combines numerical evaluation and symbolic manipulation of these expressions as follows. We assume that we have numerical values for all the quantities in matrices **A** and **b** except for K_1, K_2, $w_1(\eta_1)$, and $w_2(\xi_2)$, which are retained as symbolic quantities. Then **A** consists of only numerical values, and the elements of **b** have the form

$$b_j = n_{j1} + n_{j2} K_1 w_1(\eta_1) + n_{j3} K_2 w_2(\xi_1) \qquad j = 1, 2, 3, 4 \tag{3.29}$$

where n_{ji}, $i = 1, 2, 3$, are numerical values.

It can be shown that when the solution to Eq. (3.26) has the elements of **b** in the form given by Eq. (3.29), the solution constants c_{11}, c_{12}, c_{21}, and c_{22}, are of the form

$$\begin{aligned} c_{11} &= \frac{1}{D_s}\Big(N_{11} + N_{12} K_1 w_1(\eta_1) + N_{13} K_2 w_2(\xi_2)\Big) \\ c_{12} &= \frac{1}{D_s}\Big(N_{21} + N_{22} K_1 w_1(\eta_1) + N_{23} K_2 w_2(\xi_2)\Big) \\ c_{21} &= \frac{1}{D_s}\Big(N_{31} + N_{32} K_1 w_1(\eta_1) + N_{33} K_2 w_2(\xi_2)\Big) \\ c_{22} &= \frac{1}{D_s}\Big(N_{41} + N_{42} K_1 w_1(\eta_1) + N_{43} K_2 w_2(\xi_2)\Big) \end{aligned} \tag{3.30}$$

where N_{ij} are numerical values and $D_s = |\mathbf{A}|$, which is the determinant of the coefficients of the system of equations. It is noted that N_{11}, N_{21}, N_{31}, and N_{41} are functions of H_1 and

$\hat{H}_2$; therefore, when no loads are applied, these quantities are zero. This is important to know when the buckling loads are determined.

Substituting Eq. (3.30) into Eq. (3.25), we obtain

$$\begin{aligned} w_1(\eta) &= \frac{1}{D_s}\Big(G_{11}(\eta) + K_1 w_1(\eta_1) G_{12}(\eta,\eta_1) + K_2 w_2(\xi_2) G_{13}(\eta)\Big) \\ w_2(\xi) &= \frac{1}{D_s}\Big(\hat{G}_{21}(\xi) + K_1 w_1(\eta_1)\hat{G}_{22}(\xi) + K_2 w_2(\xi_2)\hat{G}_{23}(\xi,\xi_2)\Big) \end{aligned} \tag{3.31}$$

where

$$\begin{aligned} G_{11}(\eta) &= D_s H_1(\eta) + N_{11} p_{11}(\eta) + N_{21} p_{12}(\eta) \\ G_{12}(\eta,\eta_1) &= N_{12} p_{11}(\eta) + N_{22} p_{12}(\eta) - D_s v(\eta,\eta_1)\theta(\eta-\eta_1) \\ G_{13}(\eta) &= N_{13} p_{11}(\eta) + N_{23} p_{12}(\eta) \\ \hat{G}_{21}(\xi) &= \beta D_s \hat{H}_2(\xi) + N_{31}\hat{p}_{21}(\xi) + N_{41}\hat{p}_{22}(\xi) \\ \hat{G}_{22}(\xi) &= N_{32}\hat{p}_{21}(\xi) + N_{42}\hat{p}_{22}(\xi) \\ \hat{G}_{23}(\xi,\xi_2) &= N_{33}\hat{p}_{21}(\xi) + N_{43}\hat{p}_{22}(\xi) - \beta D_s \hat{v}(\xi,\xi_2)\theta(\xi-\xi_2) \end{aligned} \tag{3.32}$$

It is seen that $G_{11}(\xi))$ is a function of H_1 and $\hat{G}_{21}(\xi)$ is a function of $\hat{H}_2$. Since we have already determined that N_{11}, N_{21}, N_{31}, and N_{41} are also functions of H_1 and $\hat{H}_2$, when no loads are applied, $G_{11}(\xi))$ and $\hat{G}_{21}(\xi)$ equal zero.

No In-Span Rigid Supports To determine the expressions for the case of no in-span rigid supports, we set $K_1 = K_2 = 0$, and Eq. (3.31) becomes

$$\begin{aligned} w_1(\eta) &= \frac{1}{D_s} G_{11}(\eta) \\ w_2(\xi) &= \frac{1}{D_s}\hat{G}_{21}(\xi) \end{aligned} \tag{3.33}$$

Rigid Support in Each Segment To determine the expressions for the case for one rigid support in each segment of the beam, we determine $w_1(\eta_1)$ and $w_2(\xi_2)$ as follows. In the first equation of Eq. (3.31), we set $\eta = \eta_1$, and in the second equation of Eq. (3.31), we set $\xi = \xi_2$. Then, Eq. (3.31) becomes

$$\begin{aligned} w_1(\eta_1) &= \frac{1}{D_s}\Big(G_{11}(\eta_1) + K_1 w_1(\eta_1) G_{12}(\eta_1,\eta_1) + K_2 w_2(\xi_2) G_{13}(\eta_1)\Big) \\ w_2(\xi_2) &= \frac{1}{D_s}\Big(\hat{G}_{21}(\xi_2) + K_1 w_1(\eta_1)\hat{G}_{22}(\xi_2) + K_2 w_2(\xi_2)\hat{G}_{23}(\xi_2,\xi_2)\Big) \end{aligned} \tag{3.34}$$

Solving Eq. (3.34) for $w_1(\eta_1)$ and $w_2(\xi_2)$ yields

$$
\begin{aligned}
w_1(\eta_1) &= \frac{1}{D_{12}(\eta_1, \xi_2)}\Big(D_s G_{11}(\eta_1) + E_{s,1}(\eta_1, \xi_2) K_2\Big) \\
w_2(\xi_2) &= \frac{1}{D_{12}(\eta_1, \xi_2)}\Big(D_s \hat{G}_{21}(\xi_2) + E_{s,2}(\eta_1, \xi_2) K_1\Big)
\end{aligned}
\tag{3.35}
$$

where

$$
\begin{aligned}
D_{12}(\eta_1, \xi_2) &= D_s^2 - D_s G_{12}(\eta_1, \eta_1) K_1 - D_s \hat{G}_{23}(\xi_2, \xi_2) K_2 \\
&\quad + E_{s,3}(\eta_1, \xi_2) K_1 K_2 \\
E_{s,1}(\eta_1, \xi_2) &= G_{13}(\eta_1)\hat{G}_{21}(\xi_2) - G_{11}(\eta_1)\hat{G}_{23}(\xi_2, \xi_2) \\
E_{s,2}(\eta_1, \xi_2) &= \hat{G}_{22}(\xi_2) G_{11}(\eta_1) - \hat{G}_{21}(\xi_2) G_{12}(\eta_1, \eta_1) \\
E_{s,3}(\eta_1, \xi_2) &= G_{12}(\eta_1, \eta_1)\hat{G}_{23}(\xi_2, \xi_2) - G_{13}(\eta_1)\hat{G}_{22}(\xi_2)
\end{aligned}
\tag{3.36}
$$

It is noted that $E_{s,1}$ and $E_{s,2}$ are functions of H_1 and H_2; therefore, when no loads are applied, these quantities equal zero.

Substituting Eq. (3.35) into Eq. (3.31) and taking the limit as $K_1 \to \infty$ and $K_2 \to \infty$, we obtain

$$
\begin{aligned}
w_1(\eta) &= \frac{1}{D_s E_{s,3}(\eta_1, \xi_2)}\Big(E_{s,3}(\eta_1, \xi_2) G_{11}(\eta) \\
&\quad + E_{s,1}(\eta_1, \xi_2) G_{12}(\eta, \eta_1) + E_{s,2}(\eta_1, \xi_2) G_{13}(\eta)\Big) \\
w_2(\xi) &= \frac{1}{D_s E_{s,3}(\eta_1, \xi_2)}\Big(E_{s,3}(\eta_1, \xi_2)\hat{G}_{21}(\xi) \\
&\quad + E_{s,1}(\eta_1, \xi_2)\hat{G}_{22}(\xi) + E_{s,2}(\eta_1, \xi_2)\hat{G}_{23}(\xi, \xi_2)\Big)
\end{aligned}
\tag{3.37}
$$

which are the displacements of each segment of the beam with one in-span rigid support in each segment. To verify this, we note that when $\eta = \eta_1$ in the first equation of Eq. (3.37), it is straightforward to show that $w_1(\eta_1) = 0$, and when $\xi = \xi_2$ in the second equation of Eq. (3.37), it is found that $w_2(\xi_2) = 0$. Thus, at each of these locations, there is a rigid support.

Rigid Support in Segment 1 Only To determine the expressions for the case when there is a rigid support in segment 1 only, we set $K_2 = 0$, and Eq. (3.31) reduces to

$$
\begin{aligned}
w_1(\eta) &= \frac{1}{D_s}\Big(G_{11}(\eta) + K_1 w_1(\eta_1) G_{12}(\eta, \eta_1)\Big) \\
w_2(\xi) &= \frac{1}{D_s}\Big(\hat{G}_{21}(\xi) + K_1 w_1(\eta_1)\hat{G}_{22}(\xi)\Big)
\end{aligned}
\tag{3.38}
$$

In the first equation of Eq. (3.38), we set $\eta = \eta_1$ and solve for $w_1(\eta_1)$, which yields

$$w_1(\eta_1) = \frac{G_{11}(\eta_1)}{D_s - K_1 G_{12}(\eta_1, \eta_1)} \tag{3.39}$$

Substituting Eq. (3.39) into Eq. (3.38) and taking the limit as $K_1 \to \infty$, we obtain

$$\begin{aligned} w_1(\eta) &= \frac{1}{D_s G_{12}(\eta_1, \eta_1)} \Big(G_{11}(\eta) G_{12}(\eta_1, \eta_1) - G_{12}(\eta, \eta_1) G_{11}(\eta_1) \Big) \\ w_2(\xi) &= \frac{1}{D_s G_{12}(\eta_1, \eta_1)} \Big(G_{12}(\eta_1, \eta_1) \hat{G}_{21}(\xi) - G_{11}(\eta_1)) \hat{G}_{22}(\xi) \Big) \end{aligned} \tag{3.40}$$

which are the displacements of each segment of the beam when segment 1 has an in-span rigid support. To verify this, we note that when $\eta = \eta_1$ in the first equation of Eq. (3.40), we find that $w_1(\eta_1) = 0$.

Rigid Support in Segment 2 Only To determine the expressions for the case where there is a rigid support in segment 2 only, we set $K_1 = 0$, and Eq. (3.31) reduces to

$$\begin{aligned} w_1(\eta) &= \frac{1}{D_s} \Big(G_{11}(\eta) + K_2 w_2(\xi_2) G_{13}(\eta) \Big) \\ w_2(\xi) &= \frac{1}{D_s} \Big(\hat{G}_{21}(\xi) + K_2 w_2(\xi_2) \hat{G}_{23}(\xi, \xi_2) \Big) \end{aligned} \tag{3.41}$$

In the second equation of Eq. (3.41), we set $\xi = \xi_2$ and solve for $w_2(\xi_2)$, which yields

$$w_2(\xi_2) = \frac{\hat{G}_{21}(\xi_2)}{D_s - K_2 \hat{G}_{23}(\xi_2, \xi_2)} \tag{3.42}$$

Substituting Eq. (3.42) into Eq. (3.41) and taking the limit as $K_2 \to \infty$, we obtain

$$\begin{aligned} w_1(\eta) &= \frac{1}{D_s \hat{G}_{23}(\xi_2, \xi_2)} \Big(G_{11}(\eta) \hat{G}_{23}(\xi_2, \xi_2) - G_{13}(\eta) \hat{G}_{21}(\xi_2) \Big) \\ w_2(\xi) &= \frac{1}{D_s \hat{G}_{23}(\xi_2, \xi_2)} \Big(\hat{G}_{23}(\xi_2, \xi_2) \hat{G}_{21}(\xi) - \hat{G}_{21}(\xi_2) \hat{G}_{23}(\xi, \xi_2) \Big) \end{aligned} \tag{3.43}$$

which are the displacements of each segment of the beam when segment 2 has an in-span rigid support. To verify this, we note that when $\xi = \xi_2$ in the second equation of Eq. (3.43), we find that $w_2(\xi_2) = 0$.

3.1.1 Buckling Loads

When S_o is compressive, the buckling loads can be obtained directly from Eqs. (3.33), (3.37), (3.40), and (3.43) in the manner discussed in Sect. 2.2.1 by recalling that $N_{11} = N_{21} = N_{31} = N_{41} = E_{s,1} = E_{s,2} = G_{11} = \hat{G}_{21} = 0$ when no loads are applied. Then, when there are no in-span rigid supports, we use Eq. (3.33) to obtain

$$D_s = 0 \tag{3.44}$$

The values of S_o that satisfy Eq. (3.44) are the buckling loads.

When there is a rigid in-span support in each segment of the beam, we obtain from Eq. (3.37) that

$$E_{s,3}(\eta_1, \xi_2) = 0 \tag{3.45}$$

The values of S_o that satisfy Eq. (3.45) are the buckling loads.

When there is a rigid in-span support in segment 1 only, we obtain from Eq. (3.40) that

$$G_{12}(\eta_1, \eta_1) = 0 \tag{3.46}$$

The values of S_o that satisfy Eq. (3.46) are the buckling loads.

When there is a rigid in-span support in segment 2 only, we obtain from Eq. (3.43) that

$$\hat{G}_{23}(\xi_2, \xi_2) = 0 \tag{3.47}$$

The values of S_o that satisfy Eq. (3.47) are the buckling loads.

It is noted from Eqs. (2.15) and (3.24) that each function comprising these buckling load equations is a function of S_o as seen from the appropriate columns of Tables 2.2, 2.3, and 2.4. Therefore, **A** is also a function of S_o and consequently so are the numerical values N_{ij}.

3.2 Numerical Results

We illustrate the methodology by determining the responses of the displacement, rotation, moment, and shear force of the beam as a function of position to various loads, boundary conditions, in-span rigid supports, and flexural rigidity ratio β. In addition, the lowest buckling load of the configuration is also determined and displayed in each figure. The results shown are all obtained from one interactive graphic, which can evaluate the response of the beam to all combinations of factors, boundary conditions, and applied loads given in this chapter.

In all figures, the image on the top left shows the locations and relative magnitudes of the applied forces and the location of the rigid supports. The image on the top right gives the lowest buckling load for the selected configuration. The middle left image shows the nondimensional displacements $w_1(\eta)$ and $w_2(\xi)$ as a function of their respective positions, and the middle right image shows the nondimensional rotations $\Phi_1(\eta)$ and $\Phi_2(\xi)$ as a function of their respective positions. The images on the lower left and on the lower right, respectively, show the nondimensional moments $M_1(\eta)$ and $M_2(\xi)$ and the nondimensional shear forces $V_1(\eta)$ and $V_2(\xi)$ as a function of their respective positions. Where color is available, segment 1 is shown in black and segment 2 in red.

For illustrative purposes, we consider a clamped-hinged beam resting on an elastic foundation with $K_f = 60$, subjected to an axial compressive force $S_o = 6$, having a nondimensional length for segment 1 of $a_s = 0.45$, and subjected to the applied loads shown in each figure.

With $\beta = 0.5$, we display the effects of in-span rigid supports in Figs. 3.2, 3.3, 3.4, and 3.5, respectively, for no rigid supports, one rigid support in segment 1, one rigid support in segment 2, and two rigid supports, one in each segment. With $\beta = 2.0$, we display the effects of in-span rigid supports in Figs. 3.6, 3.7, 3.8, and 3.9, respectively, for no rigid supports, one rigid support in segment 1, one rigid support in segment 2, and two rigid supports, one in each segment.

From a qualitative examination of these figures, we find the following: (a) the boundary conditions are satisfied; (b) the displacements under the rigid supports are zero; (c) the displacement as a function of position is continuous; (d) the moment distribution shows an abrupt change in magnitude at those locations where an external moment has been applied; (e) the shear force distribution shows an abrupt change in magnitude at those locations where a concentrated force has been applied and where there is a rigid support; and (f) the continuity conditions where the two segments meet are satisfied.

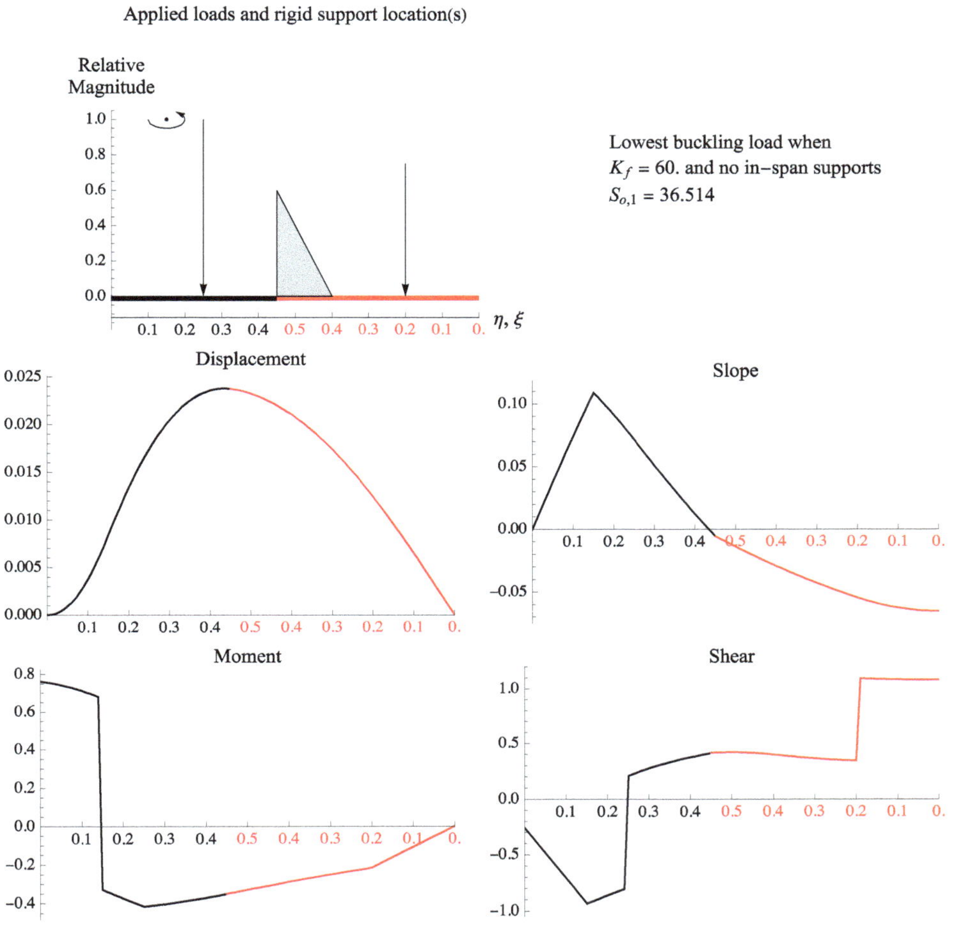

Fig. 3.2 A clamped-hinged beam with a step change in flexural rigidity of $\beta = 0.5$ at $a_s = 0.45$ with no in-span rigid supports. The beam is resting on an elastic foundation with $K_f = 60$, subjected to a compressive axial force $S_o = 6$, and subjected to the applied loads shown

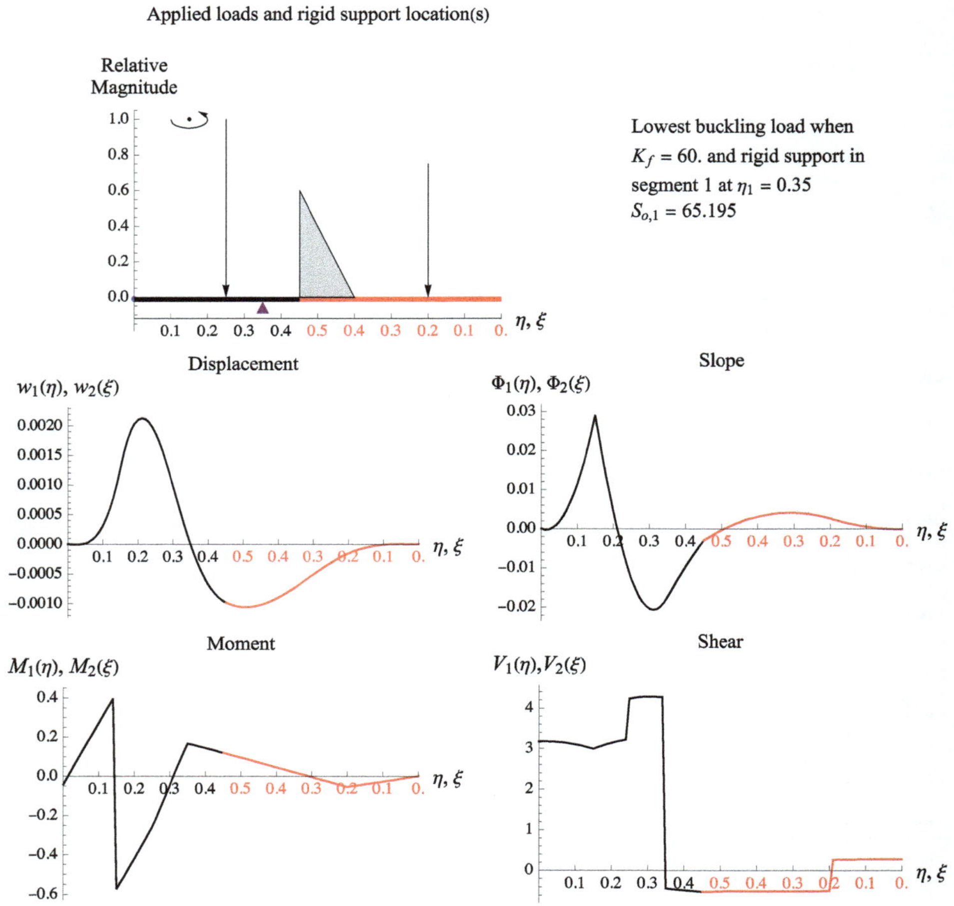

Fig. 3.3 A clamped-hinged beam with a step change in flexural rigidity of $\beta = 0.5$ at $a_s = 0.45$ with one in-span rigid support in segment 1. The beam is resting on an elastic foundation with $K_f = 60$, subjected to a compressive axial force $S_o = 6$, and subjected to the applied loads shown

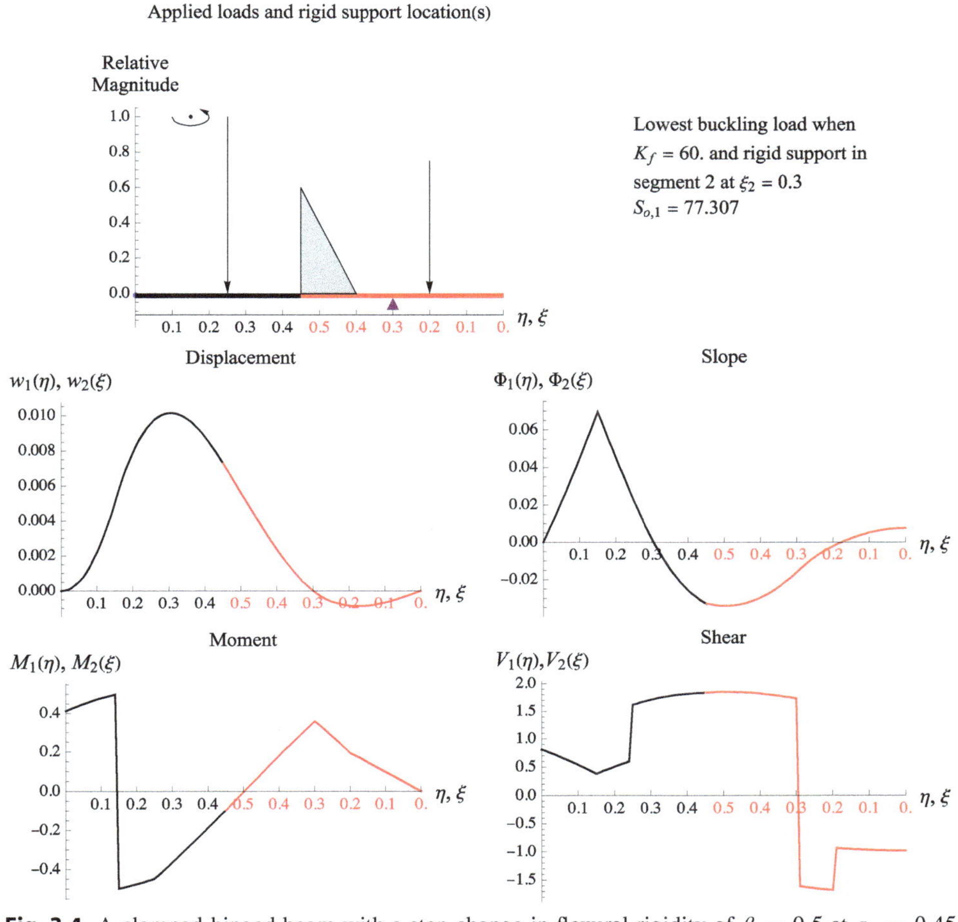

Fig. 3.4 A clamped-hinged beam with a step change in flexural rigidity of $\beta = 0.5$ at $a_s = 0.45$ with one in-span rigid support in segment 2. The beam is resting on an elastic foundation with $K_f = 60$, subjected to a compressive axial force $S_o = 6$, and subjected to the applied loads shown

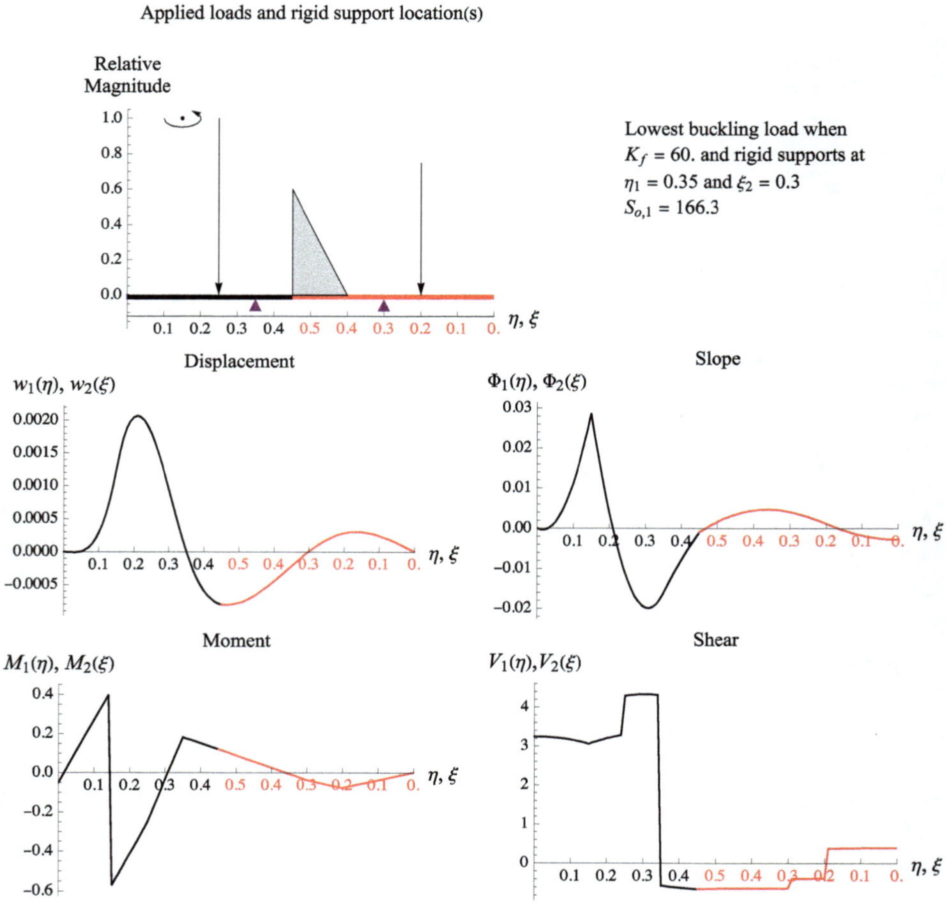

Fig. 3.5 A clamped-hinged beam with a step change in flexural rigidity of $\beta = 0.5$ at $a_s = 0.45$ with two in-span rigid supports, one in each segment. The beam is resting on an elastic foundation with $K_f = 60$, subjected to a compressive axial force $S_o = 6$, and subjected to the applied loads shown

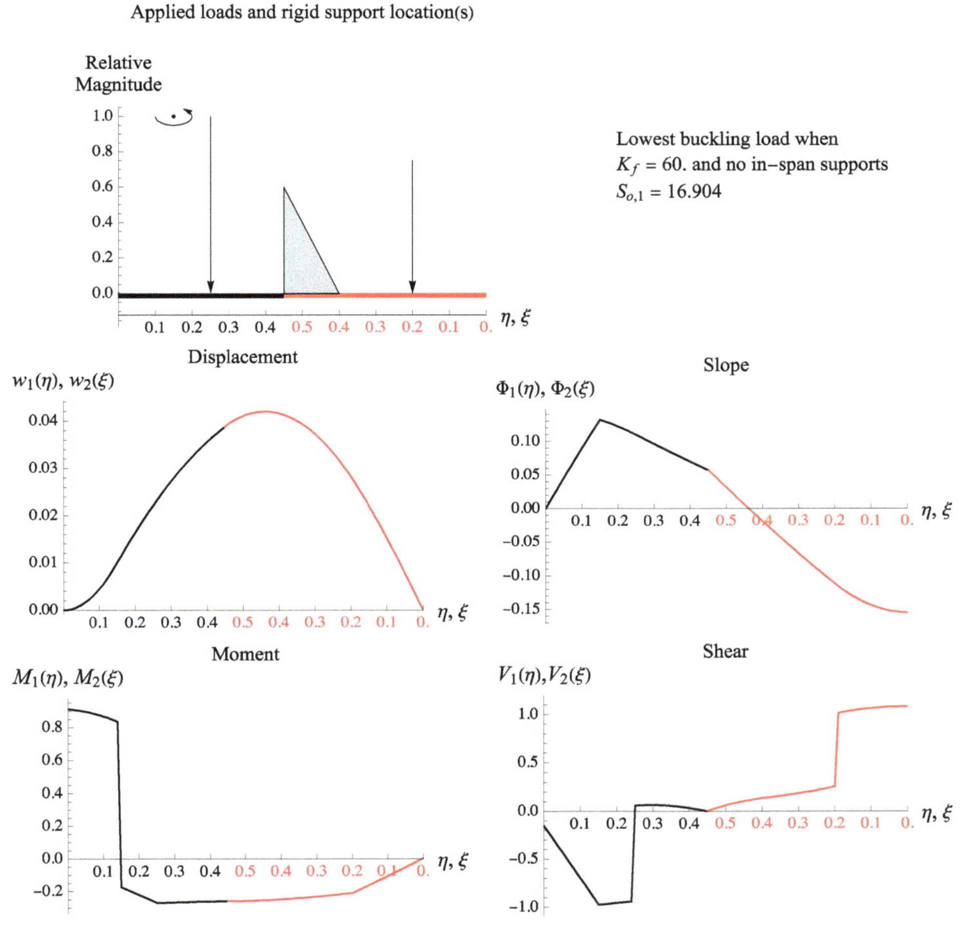

Fig. 3.6 A clamped-hinged beam with a step change in flexural rigidity of $\beta = 2.0$ at $a_s = 0.45$ with no in-span rigid supports. The beam is resting on an elastic foundation with $K_f = 60$, subjected to a compressive axial force $S_o = 6$, and subjected to the applied loads shown

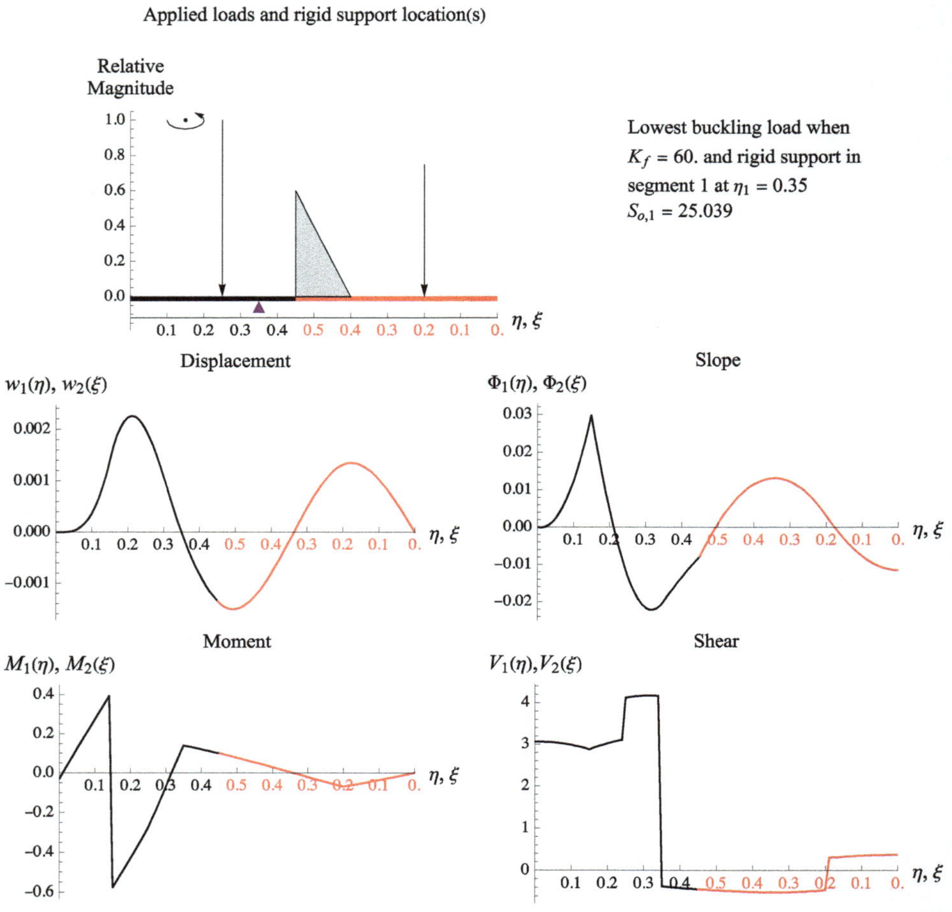

Fig. 3.7 A clamped-hinged beam with a step change in flexural rigidity of $\beta = 2.0$ at $a_s = 0.45$ with one in-span rigid support in segment 1. The beam is resting on an elastic foundation with $K_f = 60$, subjected to a compressive axial force $S_o = 6$, and subjected to the applied loads shown

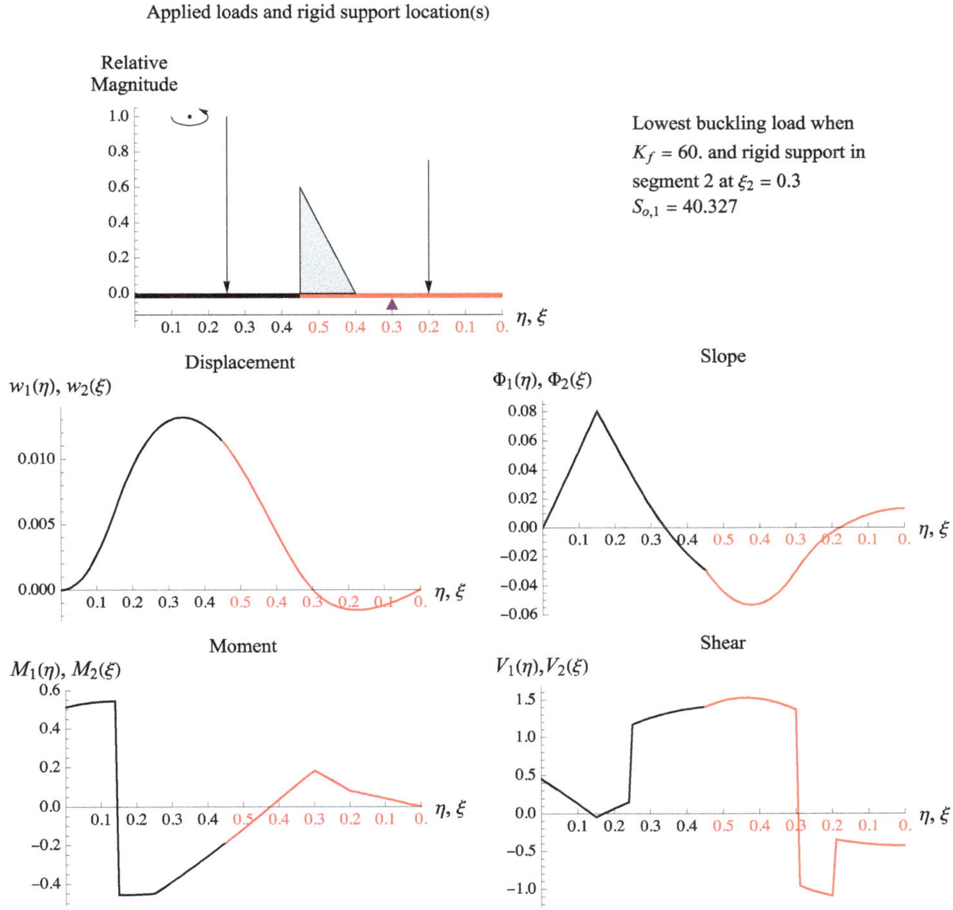

Fig. 3.8 A clamped-hinged beam with a step change in flexural rigidity of $\beta = 2.0$ at $a_s = 0.45$ with one in-span rigid support in segment 2. The beam is resting on an elastic foundation with $K_f = 60$, subjected to a compressive axial force $S_o = 6$, and subjected to the applied loads shown

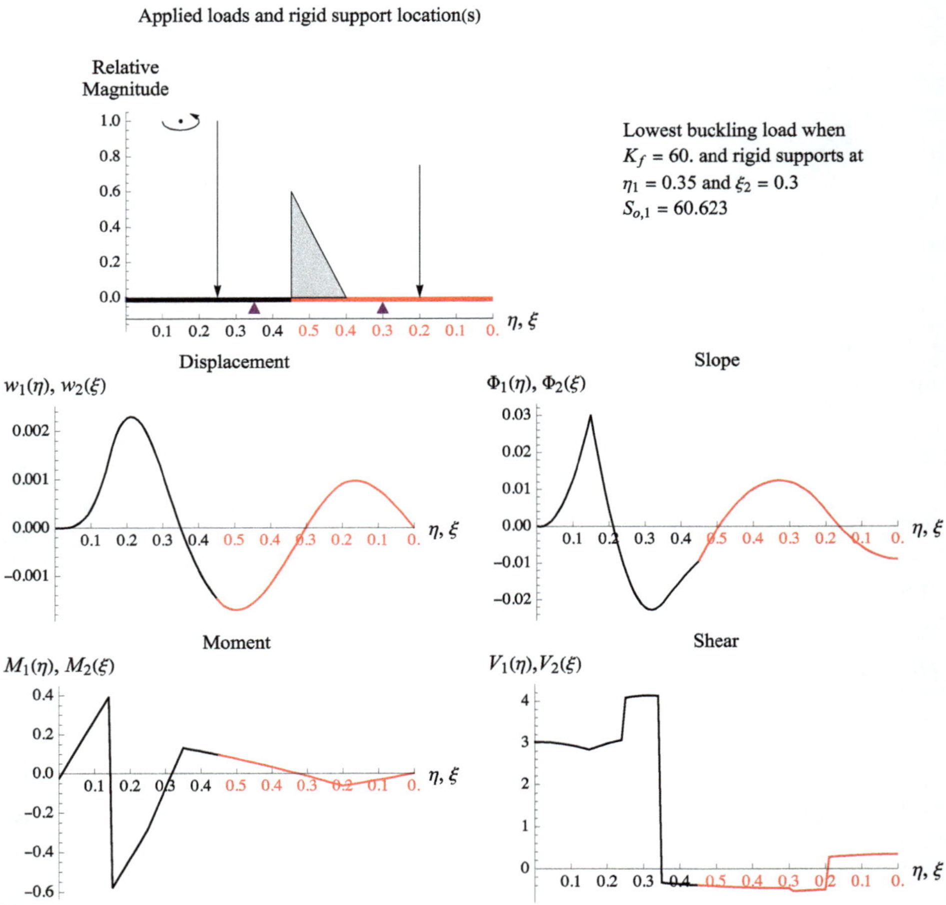

Fig. 3.9 A clamped-hinged beam with a step change in flexural rigidity of $\beta = 2.0$ at $a_s = 0.45$ with two in-span rigid supports, one in each segment. The beam is resting on an elastic foundation with $K_f = 60$, subjected to a compressive axial force $S_o = 6$, and subjected to the applied loads shown

4 Rectangular Frames

Abstract

Expressions for the spatial distribution of the displacement and moment of each member of a three-beam rectangular frame are obtained when, simultaneously to each frame member, a concentrated force and a distributed force are applied. The two grounded ends of the frame are any combination of clamped or hinged. In addition, the results are numerically extended to include a two-beam frame with each of its grounded ends fixed. The expressions for the response of the frame are obtained by using the solutions given in Chap. 2. The three-beam frame procedure is then extended to a five-beam frame and a six-beam frame.

Keywords

Frame · Beam · Concentrated force · Distributed force · Moment · Boundary conditions · Clamped · Displacement · Shear force · Slope

4.1 Three-Beam Frame: Governing Equations

Consider the three-beam rectangular frame shown in Fig. 4.1, where E_j (N/m), $j = 1, 2, 3$ is the tensile modulus, I_j (m^4) is the moment of inertia of the cross section, and L_j (m) is the length where $L_1 \geq L_2$. The sign convention used to represent the displacement y_j (m), moment M_j (Nm) at each end of the beam, and shear force V_j (N) at each end of the beam is shown in Fig. 4.2. Each beam is subjected to a trapezoidal load and a concentrated load

Supplementary Information The online version contains supplementary material available at (https://doi.org/10.1007/978-3-032-28726-7_4).

E. B. Magrab, *Static Analysis of Beams, Frames, and Arches*, Synthesis Lectures on Mechanical Engineering, https://doi.org/10.1007/978-3-032-28726-7_4

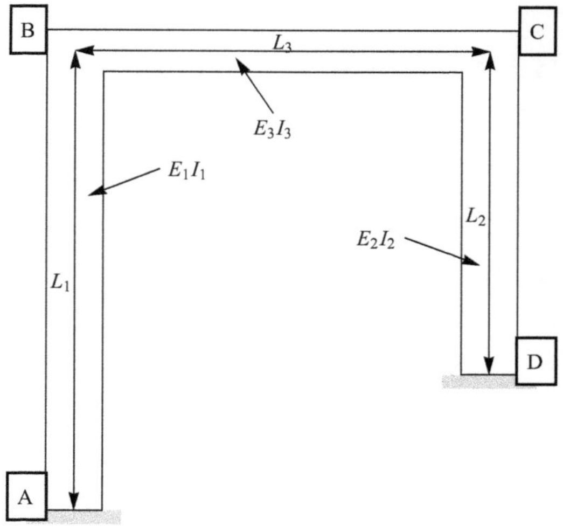

Fig. 4.1 Nomenclature for a three-beam rectangular frame

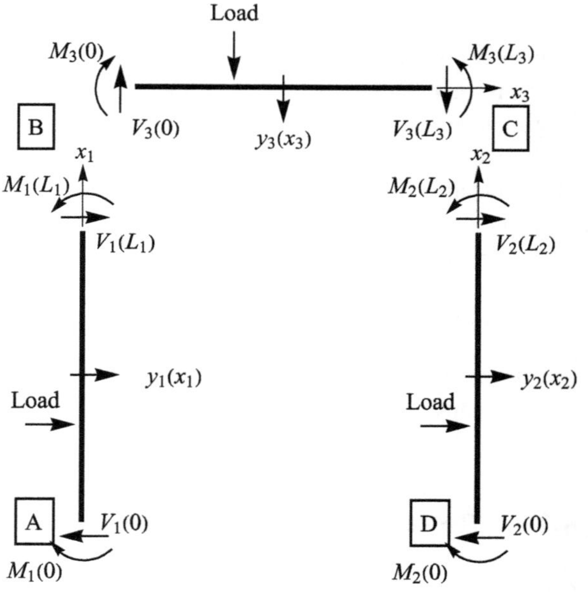

Fig. 4.2 Sign convention for the displacement, moment, shear force, and applied forces for each beam comprising the frame

as shown in Fig. 4.3. In Fig. 4.3, the quantity $F_{c,j}$ (N) is the magnitude of a concentrated force located at $L_{c,j}$ (m), $0 < L_{c,j} \le L_j$, and the quantities $F_{L,j}$ (N/m) and $F_{R,j}$ (N/m), respectively, are the magnitudes of the trapezoidal load located at $L_{L,j}$ (m) and $L_{R,j}$ (m).

The governing equation for each beam is

$$E_jI_j\frac{d^4y_j}{dx_j^4} = (a_jx_j + b_j)\big(u(x_j - L_{L,j}) - u(x_j - L_{R,j})\big) + f_{c,j}\delta(x_j - L_{c,j}) \quad \text{N/m} \quad j = 1, 2, 3 \tag{4.1}$$

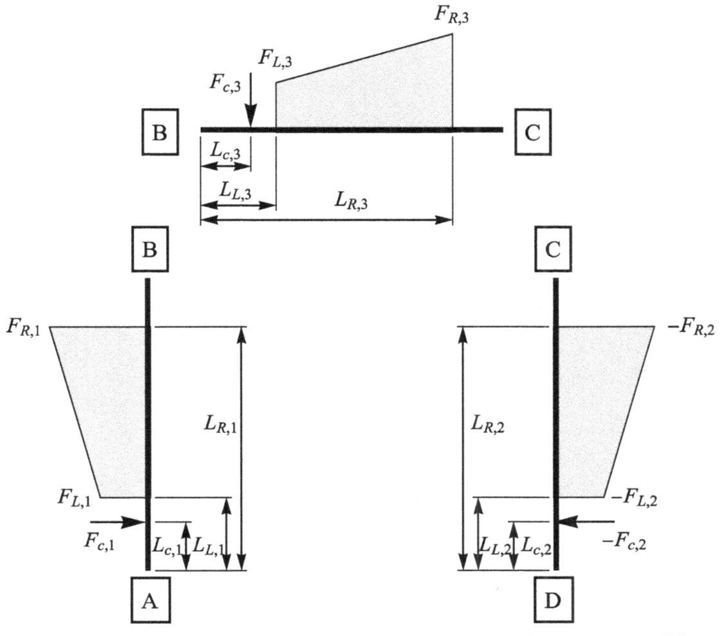

Fig. 4.3 Nomenclature for the independently applied forces to each beam comprising the frame

where

$$a_j = \frac{F_{L,j} - F_{R,j}}{L_{L,j} - L_{R,j}} \quad \text{N/m}^2$$
$$b_j = \frac{F_{R,j} L_{L,j} - F_{L,j} L_{R,j}}{L_{L,j} - L_{R,j}} \quad \text{N/m} \tag{4.2}$$

The solution to Eq. (4.1) is obtained by modifying Eq. (2.12) slightly to be

$$\begin{aligned} y_j(x_j) &= d'(x_j)y_j(0) + d(x_j)y_j'(0) + v'(x_j,0)y_j''(0) \\ &\quad + v(x_j,0)y_j'''(0) + H_j(x_j) \qquad \text{m} \end{aligned} \tag{4.3}$$

where

$$\begin{aligned} y_j(0) &= y_j(x_j)|_{x_j=0} \quad \text{m}^1 \\ y_j'(0) &= \left.\frac{dy_j(x_j)}{dx_j}\right|_{x_j=0} \quad \text{m}^0 \\ y_j''(0) &= \left.\frac{d^2y_j(x_j)}{dx_j^2}\right|_{x_j=0} \quad \text{m}^{-1} \\ y_j'''(0) &= \left.\frac{d^3y_j(x_j)}{dx_j^3}\right|_{x_j=0} \quad \text{m}^{-2} \end{aligned} \tag{4.4}$$

and

$$H_j(x_j) = \frac{1}{E_j I_j} G_j(x_j) \qquad \text{m} \tag{4.5}$$

In Eq. (4.5),

$$G_j(x_j) = a_j A_j(x_j) + b_j B_j(x_j) + F_{c,j} v(x_j, L_{c,j}) \theta(x_j - L_{c,j}) \qquad \text{Nm}^3 \tag{4.6}$$

where

$$\begin{aligned} A_j(x_j) &= \left(L_{L,j} r'(x_j, L_{L,j}) + r(x_j, L_{L,j})\right) \theta(x_j - L_{L,j}) \\ &\quad - \left(L_{R,j} r'(x_j, L_{R,j}) + r(x_j, L_{R,j})\right) \theta(x_j - L_{R,j}) \qquad \text{m}^5 \\ B_j(x_j) &= r'(x_j, L_{L,j}) \theta(x_j - L_{L,j}) - r'(x_j, L_{R,j}) \theta(x_j - L_{R,j}) \qquad \text{m}^4 \end{aligned} \tag{4.7}$$

Since $S_o = K_f = 0$, r, v, and d are obtained from the appropriate columns of Tables 2.2, 2.3, and 2.4, respectively, with η replaced by x_j. Thus,

$$\begin{aligned} r(x_j, \alpha) &= \frac{1}{120} (x_j - \alpha)^5 \quad \text{m}^5 \\ v(x_j, \alpha) &= \frac{1}{6} (x_j - \alpha)^3 \quad \text{m}^3 \\ d(x_j) &= x_j \quad \text{m} \end{aligned} \tag{4.8}$$

The rotation of each cross section $\phi_j(x_j)$ is

$$\phi_j(x_j) = \frac{dy_j}{dx_j} \tag{4.9}$$

The moment $M_j(x_j)$ of each beam is given by

$$M_j(x_j) = E_j I_j \frac{d^2 y_j}{dx_j^2} \tag{4.10}$$

and the shear force $V_j(x_j)$ of each beam is given by

$$V_j(x_j) = E_j I_j \frac{d^3 y_j}{dx_j^3} \tag{4.11}$$

There are several assumptions that are made regarding this system:

1. For beam 1, the end at A is either clamped or hinged, and for beam 2, the end at D is either clamped or hinged.

2. Under the action of an axial force, the axial displacements of beams 1 and 2 are negligible compared to the displacements of the beams caused by the trapezoidal and concentrated loads. Therefore, the displacement of beam 3 at each of its ends is taken as zero.
3. Due to assumption 2, any movement of B and C remains on the x_3 axis and the location of the x_3 axis remains fixed.
4. In the vicinity of the joint at B, it is assumed that the rotation of beam 3 equals that of beam 1 and in the vicinity of joint C the rotation of beam 3 equals that of beam 2. Thus, in these regions, the respective beam pairs remain orthogonal to each other.
5. Under the action of an axial force on beam 3, the axial displacements are negligible compared to the displacements of the beams caused by the trapezoidal and concentrated loads. Therefore, when beam 3 moves in the x_3 direction (its axial direction), it moves as a rigid body; that is, B moves the same amount as C and in the same direction.

There are 12 solution constants that have to be determined from the boundary conditions and continuity conditions. We now use the assumptions to specify the boundary conditions on each beam and the continuity conditions at B and C. We shall consider four sets of boundary conditions, which are as follows.

Boundary Conditions 1: Clamped at A and Clamped at D

$$\begin{aligned} y_1(0) &= y_1'(0) = 0 \\ y_2(0) &= y_2'(0) = 0 \\ y_3(0) &= y_3(L_3) = 0 \end{aligned} \tag{4.12}$$

With these boundary conditions, Eq. (4.3) simplifies to the following three equations:

$$\begin{aligned} y_1(x_1) &= v'(x_1, 0)y_1''(0) + v(x_1, 0)y_1'''(0) + H_1(x_1) \\ y_2(x_2) &= v'(x_2, 0)y_2''(0) + v(x_2, 0)y_2'''(0) + H_2(x_2) \\ y_3(x_3) &= d(x_3)y_3'(0) + v'(x_3, 0)y_3''(0) + v(x_3, 0)y_3'''(0) + H_3(x_3) \end{aligned} \tag{4.13}$$

Boundary Conditions 2: Hinged at A and Hinged at D

$$\begin{aligned} y_1(0) &= y_1''(0) = 0 \\ y_2(0) &= y_2''(0) = 0 \\ y_3(0) &= y_3(L_3) = 0 \end{aligned} \tag{4.14}$$

With these boundary conditions, Eq. (4.3) simplifies to the following three equations:

$$\begin{aligned}
y_1(x_1) &= d(x_1)y_1'(0) + v(x_1, 0)y_1'''(0) + H_1(x_1) \\
y_2(x_2) &= d(x_2)y_2'(0) + v(x_2, 0)y_2'''(0) + H_2(x_2) \\
y_3(x_3) &= d(x_3)y_3'(0) + v'(x_3, 0)y_3''(0) + v(x_3, 0)y_3'''(0) + H_3(x_3)
\end{aligned} \tag{4.15}$$

Boundary Conditions 3: Clamped at A and Hinged at D

$$\begin{aligned}
y_1(0) &= y_1'(0) = 0 \\
y_2(0) &= y_2''(0) = 0 \\
y_3(0) &= y_3(L_3) = 0
\end{aligned} \tag{4.16}$$

With these boundary conditions, Eq. (4.3) simplifies to the following three equations:

$$\begin{aligned}
y_1(x_1) &= v'(x_1, 0)y_1''(0) + v(x_1, 0)y_1'''(0) + H_1(x_1) \\
y_2(x_2) &= d(x_2)y_2'(0) + v(x_2, 0)y_2'''(0) + H_2(x_2) \\
y_3(x_3) &= d(x_3)y_3'(0) + v'(x_3, 0)y_3''(0) + v(x_3, 0)y_3'''(0) + H_3(x_3)
\end{aligned} \tag{4.17}$$

Boundary Conditions 4: Hinged at A and Clamped at D

$$\begin{aligned}
y_1(0) &= y_1''(0) = 0 \\
y_2(0) &= y_2'(0) = 0 \\
y_3(0) &= y_3(L_3) = 0
\end{aligned} \tag{4.18}$$

With these boundary conditions, Eq. (4.3) simplifies to the following three equations:

$$\begin{aligned}
y_1(x_1) &= d(x_1)y_1'(0)) + v(x_1, 0)y_1'''(0) + H_1(x_1) \\
y_2(x_2) &= v'(x_2, 0)y_2''(0) + v(x_2, 0)y_2'''(0) + H_2(x_2) \\
y_3(x_3) &= d(x_3)y_3'(0) + v'(x_3, 0)y_3''(0) + v(x_3, 0)y_3'''(0) + H_3(x_3)
\end{aligned} \tag{4.19}$$

Continuity Conditions Using the notation of Eqs. (4.9)–(4.11) and the directions indicated in Fig. 4.2, the six continuity conditions are as follows:

$$
\begin{aligned}
\phi_1(L_1) &= \phi_3(0) \\
\phi_2(L_2) &= \phi_3(L_3) \\
M_1(L_1) &= M_3(0) \\
M_2(L_2) &= -M_3(L_3) \\
y_1(L_1) &= y_2(L_2) \\
V_1(L_1) &= -V_2(L_2)
\end{aligned}
\tag{4.20}
$$

There are seven solution constants that are to be determined, one from the boundary condition $y_3(L_3) = 0$ and six from the continuity conditions given in Eq. (4.20). The substitution of Eqs. (4.13), (4.15), (4.17), and (4.19), respectively, into these seven conditions and using Eq. (4.8) yields the following systems of equations for the four sets of boundary conditions. It is noted that column 1 of matrix **A** given in Eq. (4.22) changes when the boundary condition of the left vertical beam goes from clamped to hinged and that column 3 changes when the boundary condition of the right vertical beam goes from clamped to hinged. The five remaining columns are unchanged as is the column matrix **b**.

Boundary Conditions 1: Clamped at A and Clamped at D

$$
\mathbf{A}_1 \mathbf{c}_1 = \mathbf{b} \tag{4.21}
$$

where

$$
\mathbf{A}_1 = \begin{pmatrix}
0 & 0 & 0 & 0 & L_3 & L_3^2/2 & L_3^3/6 \\
0 & 0 & L_2 & L_2^2/2 & -1 & -L_3 & -L_3^2/2 \\
L_1 & L_1^2/2 & 0 & 0 & -1 & 0 & 0 \\
E_1 I_1 & E_1 I_1 L_1 & 0 & 0 & 0 & -E_3 I_3 & 0 \\
0 & 0 & E_2 I_2 & E_2 I_2 L_2 & 0 & E_3 I_3 & E_3 I_3 L_3 \\
L_1^2/2 & L_1^3/6 & -L_2^2/2 & -L_2^3/6 & 0 & 0 & 0 \\
0 & E_1 I_1 & 0 & E_2 I_2 & 0 & 0 & 0
\end{pmatrix} \tag{4.22}
$$

and

$$
\mathbf{c}_1 = \begin{Bmatrix}
y_1''(0) \\ y_1'''(0) \\ y_2''(0) \\ y_2'''(0) \\ y_3'(0) \\ y_3''(0) \\ y_3'''(0)
\end{Bmatrix}
\qquad
\mathbf{b} = \begin{Bmatrix}
-H_3(L_3) \\
H_3'(L_3) - H_2'(L_2) \\
H_3'(0) - H_1'(L_1) \\
G_3''(0) - G_1''(L_1) \\
-G_3''(L_3) - G_2''(L_2) \\
H_2(L_2) - H_1(L_1) \\
-G_1'''(L_1) - G_2'''(L_2)
\end{Bmatrix} \tag{4.23}
$$

Boundary Conditions 2: Hinged at A and Hinged at D

$$\mathbf{A}_2\mathbf{c}_2 = \mathbf{b} \tag{4.24}$$

where

$$\mathbf{A}_2 = \begin{pmatrix} 0 & 0 & 0 & 0 & L_3 & L_3^2/2 & L_3^3/6 \\ 0 & 0 & 1 & L_2^2/2 & -1 & -L_3 & -L_3^2/2 \\ 1 & L_1^2/2 & 0 & 0 & -1 & 0 & 0 \\ 0 & E_1I_1L_1 & 0 & 0 & 0 & -E_3I_3 & 0 \\ 0 & 0 & 0 & E_2I_2L_2 & 0 & E_3I_3 & E_3I_3L_3 \\ L_1 & L_1^3/6 & -L_2 & -L_2^3/6 & 0 & 0 & 0 \\ 0 & E_1I_1 & 0 & E_2I_2 & 0 & 0 & 0 \end{pmatrix} \tag{4.25}$$

and

$$\mathbf{c}_2 = \left\{ y_1'(0)\ y_1'''(0)\ y_2'(0)\ y_2'''(0)\ y_3'(0)\ y_3''(0)\ y_3'''(0) \right\}^T \tag{4.26}$$

where the superscript T indicates the transpose of the row matrix.

Boundary Conditions 3: Clamped at A and Hinged at D

$$\mathbf{A}_3\mathbf{c}_3 = \mathbf{b} \tag{4.27}$$

where

$$\mathbf{A}_3 = \begin{pmatrix} 0 & 0 & 0 & 0 & L_3 & L_3^2/2 & L_3^3/6 \\ 0 & 0 & 1 & L_2^2/2 & -1 & -L_3 & -L_3^2/2 \\ L_1 & L_1^2/2 & 0 & 0 & -1 & 0 & 0 \\ E_1I_1 & E_1I_1L_1 & 0 & 0 & 0 & -E_3I_3 & 0 \\ 0 & 0 & 0 & E_2I_2L_2 & 0 & E_3I_3 & E_3I_3L_3 \\ L_1^2/2 & L_1^3/6 & -L_2 & -L_2^3/6 & 0 & 0 & 0 \\ 0 & E_1I_1 & 0 & E_2I_2 & 0 & 0 & 0 \end{pmatrix} \tag{4.28}$$

and

$$\mathbf{c}_3 = \left\{ y_1''(0)\ y_1'''(0)\ y_2'(0)\ y_2'''(0)\ y_3'(0)\ y_3''(0)\ y_3'''(0) \right\}^T \tag{4.29}$$

Boundary Conditions 4: Hinged at A and Clamped at D

$$\mathbf{A}_4\mathbf{c}_4 = \mathbf{b} \tag{4.30}$$

where

$$\mathbf{A}_4 = \begin{pmatrix} 0 & 0 & 0 & 0 & L_3 & L_3^2/2 & L_3^3/6 \\ 0 & 0 & L_2 & L_2^2/2 & -1 & -L_3 & -L_3^2/2 \\ 1 & L_1^2/2 & 0 & 0 & -1 & 0 & 0 \\ 0 & E_1I_1L_1 & 0 & 0 & 0 & -E_3I_3 & 0 \\ 0 & 0 & E_2I_2 & E_2I_2L_2 & 0 & E_3I_3 & E_3I_3L_3 \\ L_1 & L_1^3/6 & -L_2^2/2 & -L_2^3/6 & 0 & 0 & 0 \\ 0 & E_1I_1 & 0 & E_2I_2 & 0 & 0 & 0 \end{pmatrix} \tag{4.31}$$

and

$$\mathbf{c}_4 = \left\{ y_1'(0)\ y_1'''(0)\ y_2''(0)\ y_2'''(0)\ y_3'(0)\ y_3''(0)\ y_3'''(0) \right\}^T \tag{4.32}$$

The numerical solution to Eq. (4.21) is substituted into Eq. (4.13), which, in turn, is used in Eqs. (4.9), (4.10), and (4.11), respectively, to determine the rotation, moment, and shear force throughout the frame. Similarly, the numerical solution to Eq. (4.24) is substituted into Eq. (4.15), that for Eq. (4.27) is substituted into Eq. (4.17), and that for Eq. (4.30) is substituted into Eq. (4.19) and their respective results used in Eqs. (4.9), (4.10), and (4.11).

4.2 Numerical Results

The methodology is illustrated in Figs. 4.4, 4.5, 4.6, 4.7, and 4.8. It is noted that in these figures, the results are given in terms of engineering units since nondimensional quantities were not used. The values used for the applied forces, lengths, and flexural rigidities to obtain these figures are given in Table 4.1.

The image in the top center shows the locations and relative magnitudes of the forces applied to each member of the frame. The lengths of the arrows represent the relative magnitude of the concentrated forces, and the heights of the distributed loads represent their respective relative magnitudes. Two independent scales were employed since the units of the concentrated forces are different from those of the distributed loads: Newtons vs. Newtons/meter.

In the table in the center of the figure, the displacements $y_j(x_j)$, rotations $\phi_j(x_j)$, moments $M_j(x_j)$, and shear forces $V_j(x_j)$ at the ends of each beam are given. In the image on the lower left, the moment distribution along each beam comprising the frame is shown, and in the image on the lower right, the displacement of the frame is shown.

Applied loads

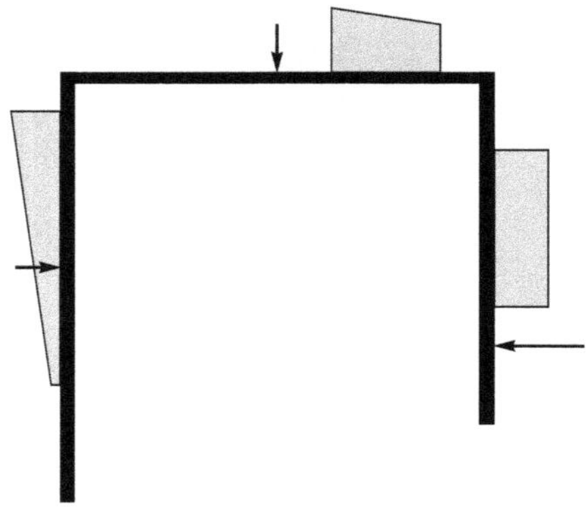

Displacement, slope, moment, and shear force at the ends of each beam

	$x_1 = 0$	$x_1 = L_1$	$x_3 = 0$	$x_3 = L_3$	$x_2 = 0$	$x_2 = L_2$
$y_j(x_j)$ (m)	0	1.901×10^{-3}	0	0	0	1.901×10^{-3}
$\phi_j(x_j)$	0.	-111.15×10^{-6}	-111.15×10^{-6}	163.8×10^{-6}	0.	163.8×10^{-6}
$M_j(x_j)$ (Nm)	4.8849×10^3	2.3899×10^3	2.3899×10^3	4.0975×10^3	-3.6274×10^3	-4.0975×10^3
$V_j(x_j)$ (N)	-1.445×10^3	1.83×10^3	-984.47	2.2655×10^3	2.57×10^3	-1.83×10^3

Moment distributions

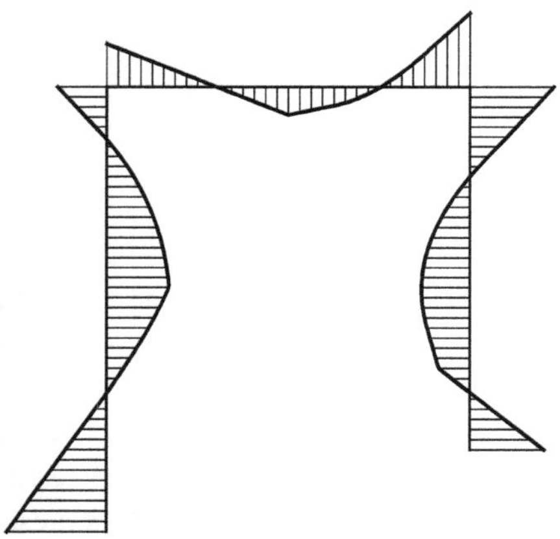

$|M_1|_{max} = 4.885 \times 10^3$ at $x_1 = 0$

$|M_2|_{max} = 4.097 \times 10^3$ at $x_2 = 9.$

$|M_3|_{max} = 4.097 \times 10^3$ at $x_3 = 8.$

Frame deflections

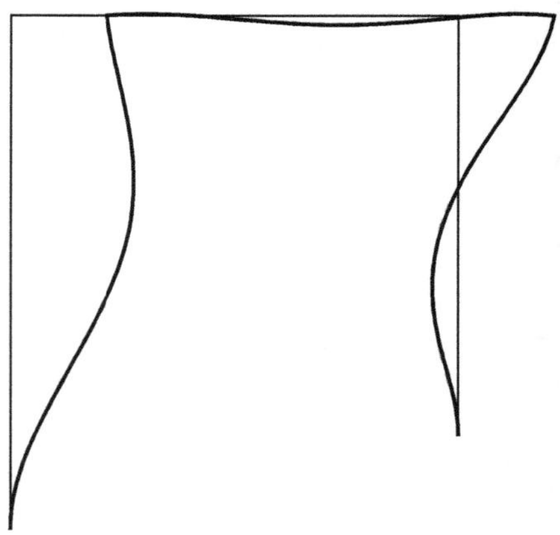

$|y_1|_{max} = 2.441 \times 10^{-3}$ at $x_1 = 7.395$

$|y_2|_{max} = 1.901 \times 10^{-3}$ at $x_2 = 8.997$

$|y_3|_{max} = 219.8 \times 10^{-6}$ at $x_3 = 4.158$

Fig. 4.4 Response of a frame to the loads shown in the figure when the left and right vertical beams are clamped. See Table 4.1 for load values. Positive values follow the directions shown in Fig. 4.2

Applied loads

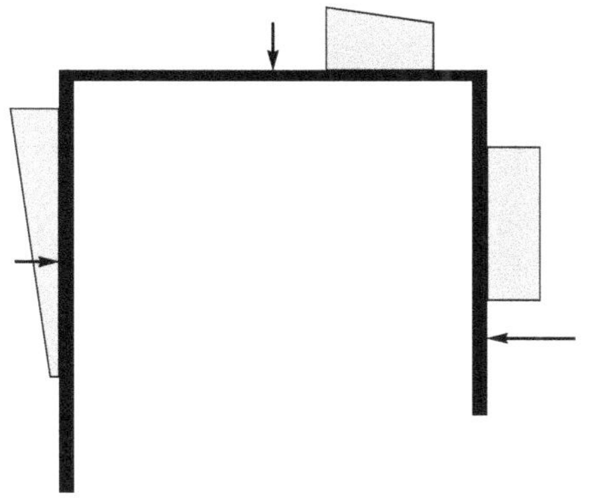

Displacement, slope, moment, and shear force at the ends of each beam

	$x_1 = 0$	$x_1 = L_1$	$x_3 = 0$	$x_3 = L_3$	$x_2 = 0$	$x_2 = L_2$
$y_j(x_j)$ (m)	0	2.704×10^{-3}	0	0	0	2.704×10^{-3}
$\phi_j(x_j)$	1.0932×10^{-3}	-399.38×10^{-6}	-399.38×10^{-6}	473.75×10^{-6}	-837.32×10^{-6}	473.75×10^{-6}
$M_j(x_j)$ (Nm)	0.	3.256×10^{3}	3.256×10^{3}	5.1754×10^{3}	0.	-5.1754×10^{3}
$V_j(x_j)$ (N)	−922.18	2.3528×10^{3}	−958.	2.292×10^{3}	2.0472×10^{3}	-2.3528×10^{3}

Moment distributions

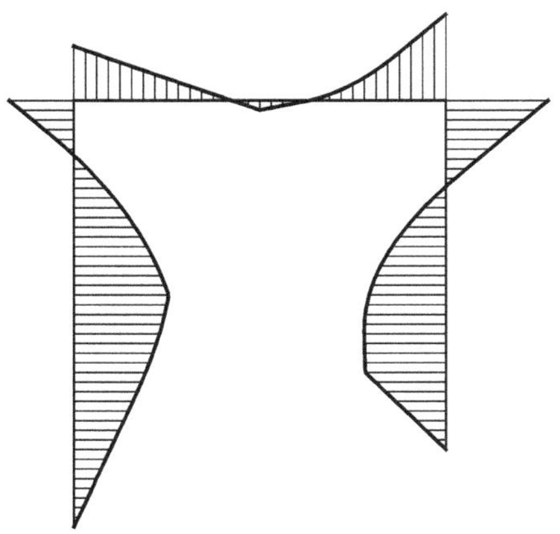

$|M_1|_{max} = 4.794 \times 10^3$ at $x_1 = 6.$

$|M_2|_{max} = 5.175 \times 10^3$ at $x_2 = 9.$

$|M_3|_{max} = 5.175 \times 10^3$ at $x_3 = 8.$

Frame deflections

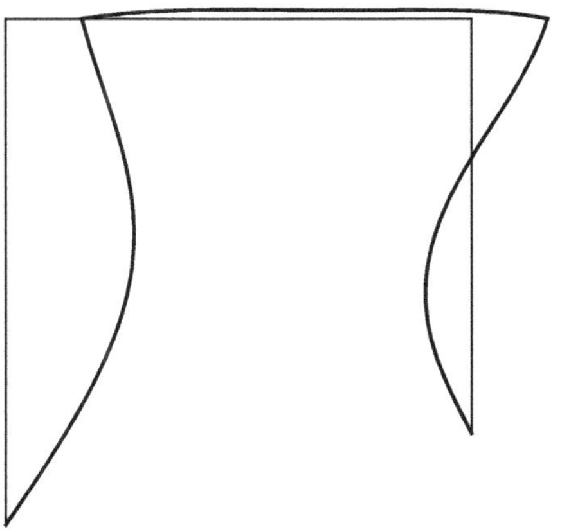

$|y_1|_{max} = 4.53 \times 10^{-3}$ at $x_1 = 6.333$

$|y_2|_{max} = 2.704 \times 10^{-3}$ at $x_2 = 9.$

$|y_3|_{max} = 407.8 \times 10^{-6}$ at $x_3 = 2.555$

Fig. 4.5 Response of a frame to the loads shown in the figure when the left and right vertical beams are hinged. See Table 4.1 for load values. Positive values follow the directions shown in Fig. 4.2

Applied loads

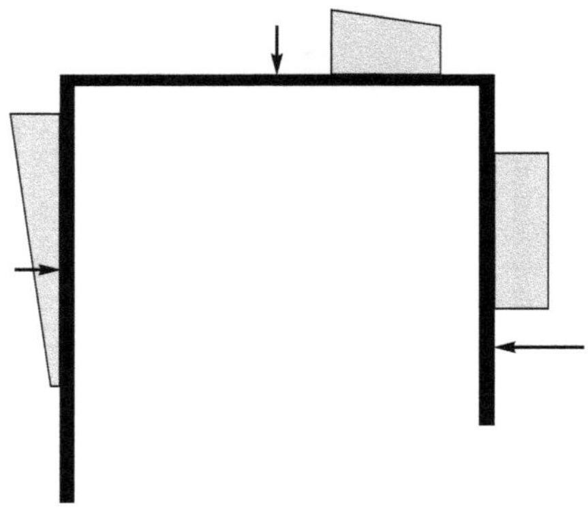

Displacement, slope, moment, and shear force at the ends of each beam

	$x_1 = 0$	$x_1 = L_1$	$x_3 = 0$	$x_3 = L_3$	$x_2 = 0$	$x_2 = L_2$
$y_j(x_j)$ (m)	0	-2.4518×10^{-3}	0	0	0	-2.4518×10^{-3}
$\phi_j(x_j)$	0.	-424.22×10^{-6}	-424.22×10^{-6}	287.77×10^{-6}	-1.6036×10^{-3}	287.77×10^{-6}
$M_j(x_j)$ (Nm)	2.3421×10^{3}	4.0219×10^{3}	4.0219×10^{3}	3.8858×10^{3}	0.	-3.8858×10^{3}
$V_j(x_j)$ (N)	-1.0655×10^{3}	2.2095×10^{3}	-1.2149×10^{3}	2.0351×10^{3}	2.1905×10^{3}	-2.2095×10^{3}

Moment distributions

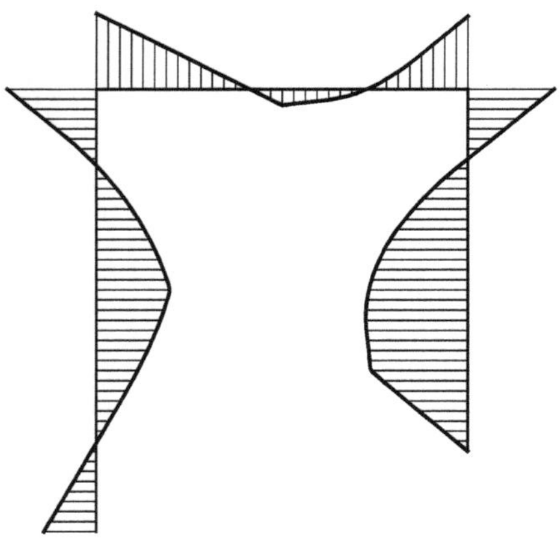

$|M_1|_{max} = 4.022\times10^{3}$ at $x_1 = 11.$
$|M_2|_{max} = 4.602\times10^{3}$ at $x_2 = 3.317$
$|M_3|_{max} = 4.022\times10^{3}$ at $x_3 = 0$

Frame deflections

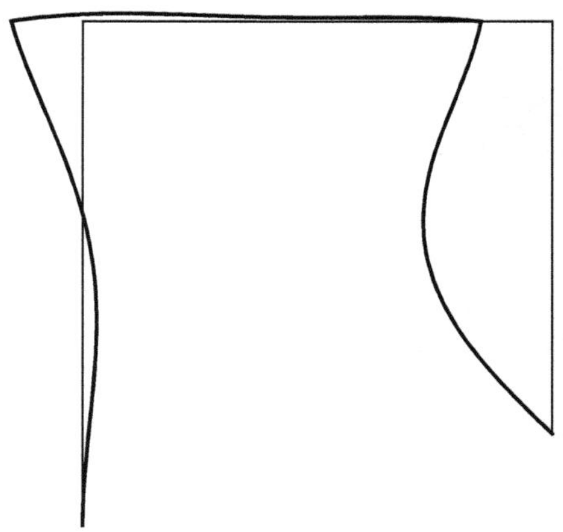

$|y_1|_{max} = 2.452\times10^{-3}$ at $x_1 = 11.$
$|y_2|_{max} = 4.425\times10^{-3}$ at $x_2 = 4.607$
$|y_3|_{max} = 354.5\times10^{-6}$ at $x_3 = 1.94$

Fig. 4.6 Response of a frame to the loads shown in the figure when the left vertical beam is clamped and the right vertical beam is hinged. See Table 4.1 for load values. Positive values follow the directions shown in Fig. 4.2

Applied loads

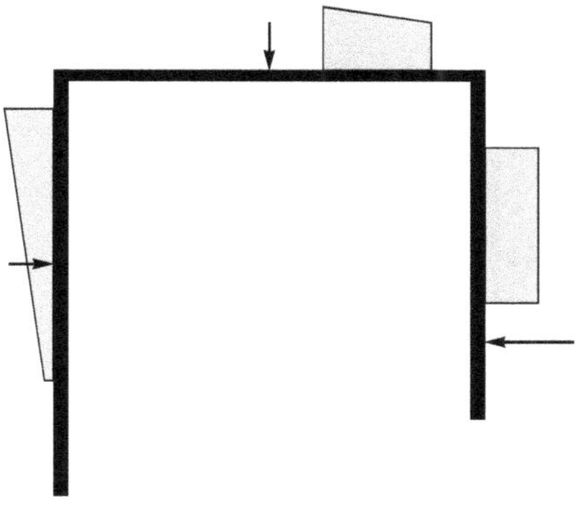

Displacement, slope, moment, and shear force at the ends of each beam

	$x_1 = 0$	$x_1 = L_1$	$x_3 = 0$	$x_3 = L_3$	$x_2 = 0$	$x_2 = L_2$
$y_j(x_j)$ (m)	0	6.1248×10^{-3}	0	0	0	6.1248×10^{-3}
$\phi_j(x_j)$	1.5012×10^{-3}	-282.49×10^{-6}	-282.49×10^{-6}	500.55×10^{-6}	0.	500.55×10^{-6}
$M_j(x_j)$ (Nm)	0.	2.4091×10^{3}	2.4091×10^{3}	5.7295×10^{3}	-1.247×10^{3}	-5.7295×10^{3}
$V_j(x_j)$ (N)	-999.17	2.2758×10^{3}	-782.87	2.4671×10^{3}	2.1242×10^{3}	-2.2758×10^{3}

Moment distributions

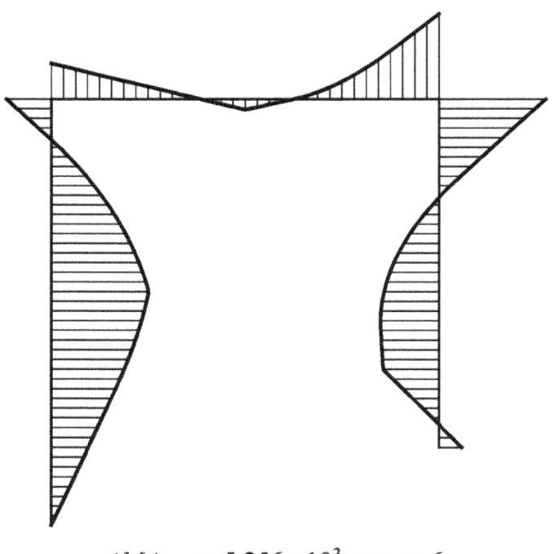

$|M_1|_{max} = 5.256\times10^3$ at $x_1 = 6.$

$|M_2|_{max} = 5.73\times10^3$ at $x_2 = 9.$

$|M_3|_{max} = 5.73\times10^3$ at $x_3 = 8.$

Frame deflections

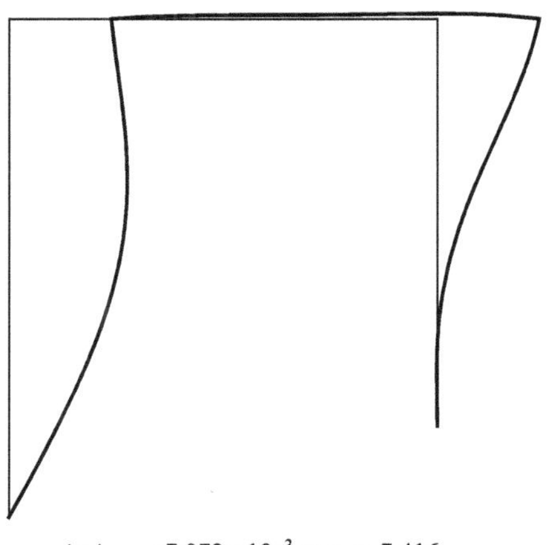

$|y_1|_{max} = 7.073\times10^{-3}$ at $x_1 = 7.416$

$|y_2|_{max} = 6.125\times10^{-3}$ at $x_2 = 9.$

$|y_3|_{max} = 370.9\times10^{-6}$ at $x_3 = 6.114$

Fig. 4.7 Response of a frame to the loads shown in the figure when the left vertical beam is hinged and the right vertical beam is clamped. See Table 4.1 for load values. Positive values follow the directions shown in Fig. 4.2

Applied loads

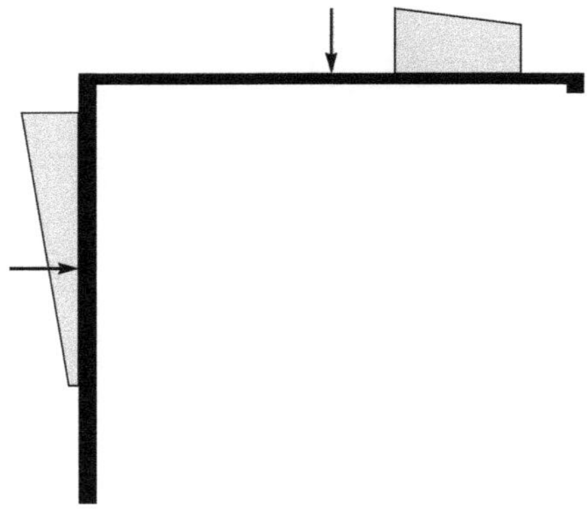

Displacement, slope, moment, and shear force at the ends of each beam

	$x_1 = 0$	$x_1 = L_1$	$x_3 = 0$	$x_3 = L_3$	$x_2 = 0$	$x_2 = L_2$
$y_j(x_j)$ (m)	0	-261.04×10^{-9}	0	0	0	-261.04×10^{-9}
$\phi_j(x_j)$	0.	-190.1×10^{-6}	-190.1×10^{-6}	-1.1292×10^{-6}	0.	-1.1292×10^{-6}
$M_j(x_j)$ (Nm)	3.6061×10^3	3.439×10^3	3.439×10^3	2.7689×10^3	-1.748×10^3	-2.7689×10^3
$V_j(x_j)$ (N)	-1.2334×10^3	2.0416×10^3	-1.2817×10^3	1.9683×10^3	-2.0416×10^3	-2.0416×10^3

Moment distributions

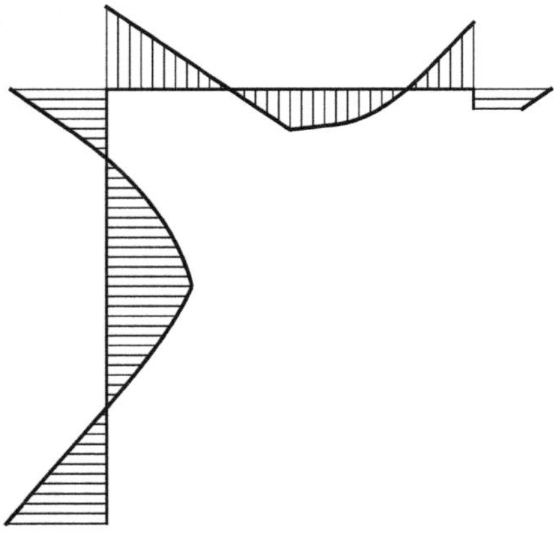

$|M_1|_{max} = 3.606\times10^3$ at $x_1 = 0$
$|M_2|_{max} = 2.769\times10^3$ at $x_2 = 0.5$
$|M_3|_{max} = 3.439\times10^3$ at $x_3 = 0$

Frame deflections

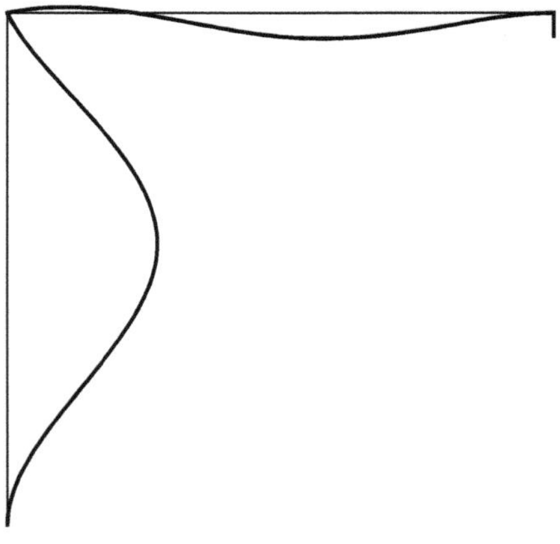

$|y_1|_{max} = 1.311\times10^{-3}$ at $x_1 = 6.034$
$|y_2|_{max} = 188.5\times10^{-9}$ at $x_2 = 0.4299$
$|y_3|_{max} = 329.2\times10^{-6}$ at $x_3 = 4.637$

Fig. 4.8 Approximate response of a two-beam frame to the loads shown in the figure when the two ends are clamped. See Table 4.1 for load values. Positive values follow the directions shown in Fig. 4.2

Table 4.1 Values used to obtain Figs. 4.4, 4.5, 4.6, 4.7, and 4.8

Parameter	$j = 1$	$j = 2$	$j = 3$
E_jI_j (Nm^2)	16.0×10^6	10.0×10^6	13.0×10^6
L_j (m)	11.0	9.0	8.0
$f_{c,j}$ (N)	1000.0	−2000.0	1500.0
$L_{c,j}$ (m)	6.0	2.0	4.0
$F_{L,j}$ (N)	550.0	−600.0	1000.0
$L_{L,j}$ (m)	10.0	3.0	5.0
$F_{R,j}$ (N)	100.0	−600.0	750.0
$L_{R,j}$ (m)	3.0	7.0	7.0

Accompanying the image of the moment distribution are the maximum absolute values of the moment in each member of the frame and their locations, and accompanying the image of the frame displacements are the maximum absolute values of the displacements in each member and their locations. The results were obtained from one interactive graphic, which can evaluate the response of the beam to a virtually unlimited combination of reasonable values for flexural rigidity, beam length, and concentrated and distributed loads.

From a qualitative examination of these figures, we find the following: (a) the boundary conditions at A and D are satisfied; (b) the moment distributions have a discontinuity where a concentrated force has been applied; (c) points B and C move the same amount and in the same direction and remain constrained to move along the axis of the beam; (d) at the connection points B and C, the respective beams remain orthogonal to each other; and (e) the magnitudes of the moments of the respective beams at B and C are equal.

Two-Beam Frame A two-beam frame can be approximated from the three-beam model as follows. For beam 2, we assume that it is clamped, remove the applied forces, set its length to 0.5 m, and set its flexural rigidity to 10^9 Nm^2. Beam 1, the remaining vertical beam, can be either clamped or hinged. Assuming that the grounded end of beam 1 is clamped and the remaining parameter values are set to those appearing in Table 4.1, we obtain Fig. 4.8.

4.3 Five-Beam Frame: Boundary and Continuity Conditions

The results of Sect. 4.1 are now extended by considering a five-beam frame shown in Fig. 4.9. There are now five governing ordinary differential equations that require the determination of 20 solution constants from the boundary conditions and continuity conditions. For specificity, we shall assume that the three vertical beams are clamped to the ground.

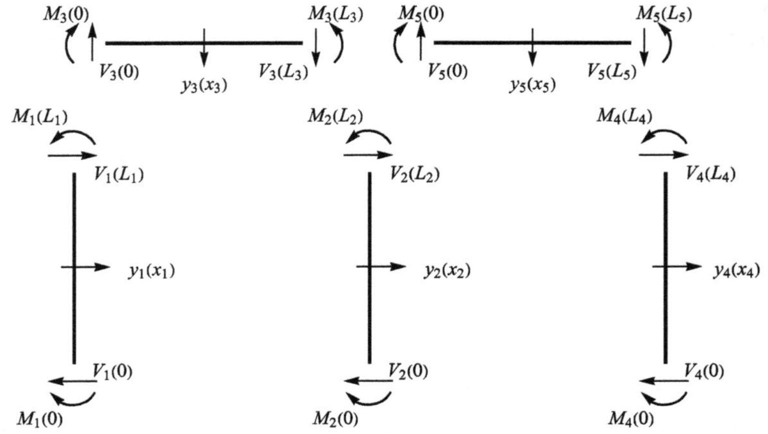

Fig. 4.9 Nomenclature for a five-beam rectangular frame

The boundary conditions are as follows:

$$\begin{aligned} y_1(0) &= y_1'(0) = 0 \\ y_2(0) &= y_2'(0) = 0 \\ y_4(0) &= y_4'(0) = 0 \\ y_3(0) &= y_3(L_3) = 0 \\ y_5(0) &= y_5(L_5) = 0 \end{aligned} \tag{4.33}$$

With these boundary conditions, Eq. (4.3) simplifies to the following five equations:

$$\begin{aligned} y_1(x_1) &= v'(x_1,0)y_1''(0) + v(x_1,0)y_1'''(0) + H_1(x_1) \\ y_2(x_2) &= v'(x_2,0)y_2''(0) + v(x_2,0)y_2'''(0) + H_2(x_2) \\ y_3(x_3) &= d(x_3)y_3'(0) + v'(x_3,0)y_3''(0) + v(x_3,0)y_3'''(0) + H_3(x_3) \\ y_4(x_4) &= v'(x_4,0)y_4''(0) + v(x_4,0)y_4'''(0) + H_4(x_4) \\ y_5(x_5) &= d(x_5)y_5'(0) + v\prime(x_5,0)y_5''(0) + v(x_5,0)y_5'''(0) + H_5(x_5) \end{aligned} \tag{4.34}$$

where $H_j(x_j)$ is given by Eq. (4.5).

Using the notation shown in Fig. 4.9, the ten continuity conditions are

$$\begin{aligned}
\phi_1(L_1) &= \phi_3(0) \\
\phi_2(L_2) &= \phi_3(L_3) \\
\phi_3(L_3) &= \phi_5(0) \\
\phi_5(L_5) &= \phi_4(L_4) \\
M_1(L_1) &= M_3(0) \\
M_3(L_3) &= -M_2(L_2) + M_5(0) \\
M_4(L_4) &= -M_5(L_5) \\
y_1(L_1) &= y_2(L_2) \\
y_2(L_2) &= y_4(L_4) \\
V_1(L_1) &= -V_2(L_2) - V_4(L_4)
\end{aligned} \tag{4.35}$$

The solution is obtained by employing the same procedure as was used to arrive at Eq. (4.21).

4.4 Six-Beam Frame: Boundary and Continuity Conditions

The results of Sect. 4.1 are now extended by considering a six-beam frame shown in Fig. 4.10. Since only rectangular frames are considered, $L_4 = L_5$ and $L_3 = L_6$. There are six governing ordinary differential equations that require the determination of 24 solution constants from the boundary conditions and continuity conditions. For specificity, we shall assume that the two lower vertical beams are clamped to the ground.

The boundary conditions are as follows:

$$\begin{aligned}
y_1(0) &= y_1'(0) = 0 \\
y_2(0) &= y_2'(0) = 0 \\
y_3(0) &= y_3(L_3) = 0 \\
y_6(0) &= y_6(L_6) = 0
\end{aligned} \tag{4.36}$$

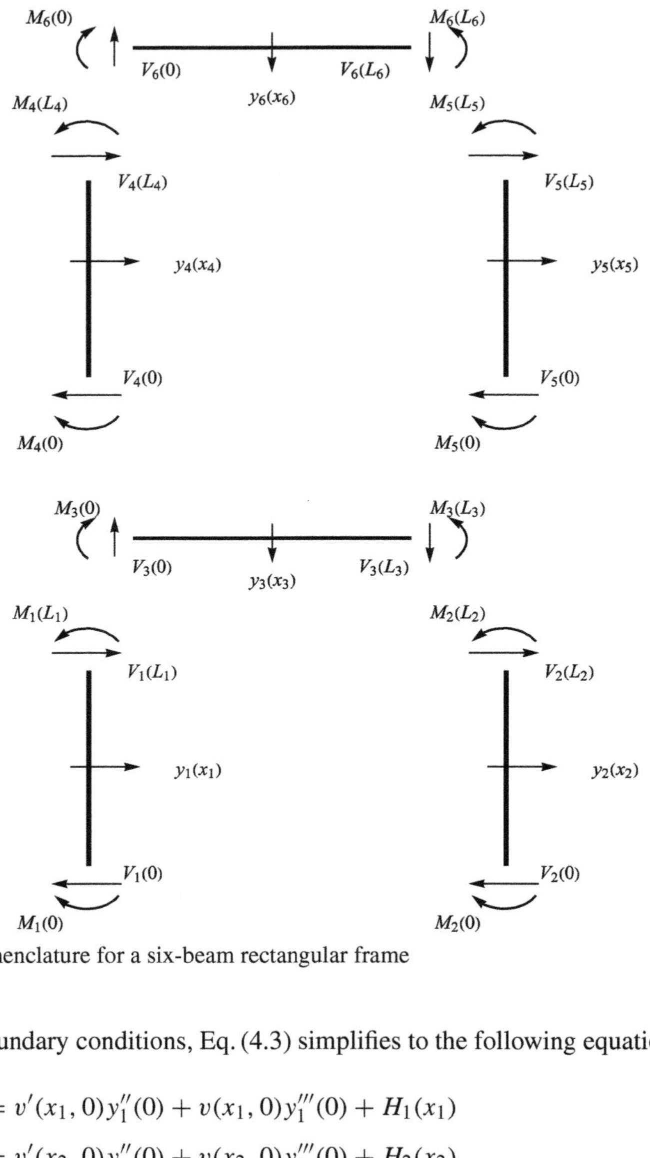

Fig. 4.10 Nomenclature for a six-beam rectangular frame

With these boundary conditions, Eq. (4.3) simplifies to the following equations:

$$
\begin{aligned}
y_1(x_1) &= v'(x_1,0)y_1''(0) + v(x_1,0)y_1'''(0) + H_1(x_1) \\
y_2(x_2) &= v'(x_2,0)y_2''(0) + v(x_2,0)y_2'''(0) + H_2(x_2) \\
y_3(x_3) &= d(x_3)y_3'(0) + v'(x_3,0)y_3''(0) + v(x_3,0)y_3'''(0) + H_3(x_3) \\
y_4(x_4) &= d'(x_4)y_4(0) + d(x_4)y_4'(0) + v'(x_4,0)y_4''(0) + v(x_4,0)y_4'''(0) \\
&\quad + H_4(x_4) \\
y_5(x_5) &= d'(x_5)y_5(0) + d(x_5)y_5'(0) + v'(x_5,0)y_5''(0) + v(x_5,0)y_5'''(0) \\
&\quad + H_5(x_5) \\
y_6(x_6) &= d(x_6)y_6'(0) + v'(x_6,0)y_6''(0) + v(x_6,0)y_6'''(0) + H_6(x_6)
\end{aligned}
\tag{4.37}
$$

where $H_j(x_j)$ is given by Eq. (4.5).

Using the notation shown in Fig. 4.10, the 16 continuity conditions are

$$
\begin{aligned}
\phi_1(L_1) &= \phi_3(0) \\
\phi_2(L_2) &= \phi_3(L_3) \\
\phi_3(L_3) &= \phi_5(0) \\
\phi_3(0) &= \phi_4(0) \\
\phi_6(0) &= \phi_4(L_4) \\
\phi_6(L_6) &= \phi_5(L_5) \\
M_1(L_1) &= M_3(0) \\
M_3(L_3 &= -M_2(L_2) \\
M_4(L_4) &= -M_6(0) \\
M_6(L_6) &= -M_5(L_5) \\
y_1(L_1) &= y_2(L_2) \\
y_4(L_4) &= y_5(L_5) \\
y_4(0) &= y_1(L_1) \\
y_5(0) &= y_2(L_2) \\
V_4(L_4) &= -V_5(L_5) \\
V_4(0) &= -V_5(0) + V_1(L_1) + V_2(L_2)
\end{aligned}
\tag{4.38}
$$

The solution is obtained by employing the same procedure as that used to arrive at Eq. (4.21).

5 Rectangular Plates with Two Opposite Edges Hinged

Abstract

In this chapter, we show that the solution obtained in Chap. 2 is also a solution to a rectangular plate hinged on two opposite edges when the plate is subjected to a trapezoidal load, resting on an elastic foundation, and subjected to two orthogonal in-plane compressive forces. The boundary conditions on the remaining two edges can be either clamped, hinged, or free.

Keywords

Plate · Hinged edge · Trapezoidal load · Boundary conditions · Clamped edge · Free edge · Shear force · Elastic foundation · In-plane force · Moment

5.1 Governing Equation

Consider the rectangular plate shown in Fig. 5.1, which has a length a_o (m) in the x-direction, a length b_o (m) in the y-direction, and a constant thickness h_p (m) in the z-direction, where $h_p << a_o$ and $h_p << b_o$. The plane of a uniformly thick plate that lies equidistant from the top and bottom surfaces is called the middle plane. The plate has a Young's modulus E (N/m^2) and a Poisson's ratio ν. We assume that the plate rests on an elastic foundation with coefficient k_f (N/m^3) and is subjected to two constant in-plane compressive forces per unit length acting on the middle plane of the plate: N_x (N/m),

Supplementary Information The online version contains supplementary material available at (https://doi.org/10.1007/978-3-032-28726-7_5).

E. B. Magrab, *Static Analysis of Beams, Frames, and Arches*, Synthesis Lectures on Mechanical Engineering, https://doi.org/10.1007/978-3-032-28726-7_5

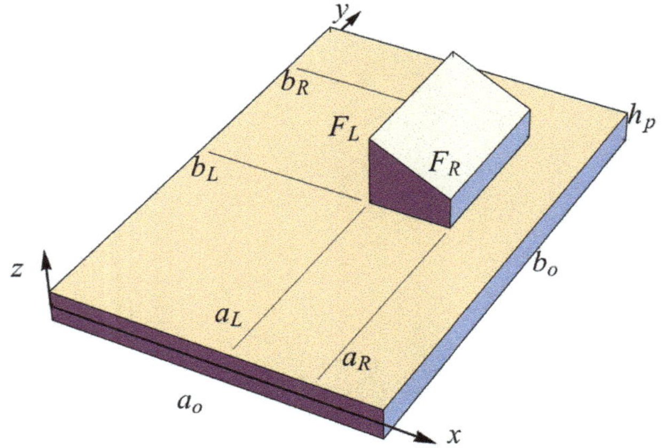

Fig. 5.1 Nomenclature for a rectangular plate and its applied trapezoidal-like load. In-plane forces are not shown

which acts in a direction parallel to the x-axis, and N_y (N/m), which acts in a direction parallel to the y-axis. In addition, we assume that the distributed force $F_a(x, y)$ (N/m^2) is normal to the surface of the plate. Then, the governing equation is given by[1]

$$D\nabla^4 w + N_x \frac{\partial^2 w}{\partial x^2} + N_y \frac{\partial^2 w}{\partial y^2} + k_f w = F_a(x, y) \tag{5.1}$$

where

$$\nabla^4 = \frac{\partial^4}{\partial x^4} + 2\frac{\partial^4}{\partial x^2 \partial y^2} + \frac{\partial^4}{\partial y^4} \tag{5.2}$$

and

$$D = \frac{E h_p^3}{12(1 - \nu^2)} \quad \text{Nm} \tag{5.3}$$

As shown in Fig. 5.1, we have assumed that the applied distributed load is constant in the y-direction and, independently, varies linearly in the x-direction. Thus, $F_a(x, y) = f_x(x) f_y(y)$, where

$$\begin{aligned} f_y(y) &= u(y - b_L) - u(y - b_R) \\ f_x(x) &= (a_p x + b_p)\big(u(x - a_L) - u(x - a_R)\big) \quad \text{N/m}^2 \end{aligned} \tag{5.4}$$

[1] [1] in Chapter 2, p.347.

$u(x)$ is the unit step function and

$$a_p = \frac{F_L - F_R}{a_L - a_R} \quad \text{N/m}^3$$
$$b_p = \frac{F_R a_L - F_L a_R}{a_L - a_R} \quad \text{N/m}^2 \tag{5.5}$$

where a_L (m) and a_R (m) are the locations that define the ends of the distributed load in the x-direction and b_L (m) and b_R (m) are the locations that define the ends of the distributed load in the y-direction. The quantities F_L (N/m^2) and F_R (N/m^2), respectively, are the magnitudes of the distributed load at a_L and a_R.

The moment per unit length about a line in the middle plane parallel to the y-axis, denoted M_x, and the moment about a line in the middle plane parallel to the x-axis, denoted M_y, respectively, are determined from

$$M_x(x, y) = -D\left(\frac{\partial^2 w}{\partial x^2} + \nu\frac{\partial^2 w}{\partial y^2}\right) \quad \text{N}$$
$$M_y(x, y) = -D\left(\frac{\partial^2 w}{\partial y^2} + \nu\frac{\partial^2 w}{\partial x^2}\right) \quad \text{N} \tag{5.6}$$

The shear force per unit length on a cross section parallel to the y-axis, denoted V_x, and the shear force on a cross section parallel to the x-axis, denoted V_y, respectively, are determined from

$$V_x(x, y) = D\left(\frac{\partial^3 w}{\partial x^3} + (2-\nu)\frac{\partial^3 w}{\partial x \partial y^2}\right) \quad \text{N/m}$$
$$V_y(x, y) = D\left(\frac{\partial^3 w}{\partial y^3} + (2-\nu)\frac{\partial^3 w}{\partial y \partial x^2}\right) \quad \text{N/m} \tag{5.7}$$

Several common boundary conditions for this system are given in Table 5.1.

Table 5.1 Several common boundary conditions for a rectangular plate: dimensional form

Boundary	$x = 0$ or a_o	$y = 0$ or b_o
Clamped	$w(x, y)\big\vert_{x=0,a_o} = 0$	$w(x, y)\big\vert_{y=0,b_o} = 0$
	$\frac{\partial w(x,y)}{\partial x}\Big\vert_{x=0,a_o} = 0$	$\frac{\partial w(x,y)}{\partial y}\Big\vert_{y=0,b_o} = 0$
Hinged	$w(x, y)\big\vert_{x=0,a_o} = 0$	$w(x, y)\big\vert_{y=0,b_o} = 0$
	$M_x(x, y)\big\vert_{x=0,a_o} = 0$	$M_y(x, y)\big\vert_{y=0,b_o} = 0$
Free	$M_x(x, y)\big\vert_{x=0,a_o} = 0$	$M_y(x, y)\big\vert_{y=0,b_o} = 0$
	$V_x(x, y)\big\vert_{x=0,a_o} + N_x \frac{\partial w(x,y)}{\partial x}\Big\vert_{x=0,a_o} = 0$	$V_y(x, y)\big\vert_{y=0,b_o} + N_y \frac{\partial w(x,y)}{\partial y}\Big\vert_{y=0,b_o} = 0$

To place the governing equations and boundary conditions into a nondimensional form, we introduce the parameters $\eta = x/a_o$, $\xi = y/a_o$, and

$$\hat{w} = \frac{Dw}{F_{max}a_o^4} \tag{5.8}$$

where F_{max} (N/m^2) is the maximum value of F_L and F_R. Upon substituting these quantities into Eqs. (5.1) and (5.2), respectively, we obtain

$$\nabla^4_{\eta\xi}\hat{w} + \hat{N}_\eta \frac{\partial^2 \hat{w}}{\partial \eta^2} + \hat{N}_\xi \frac{\partial^2 \hat{w}}{\partial \xi^2} + K_F \hat{w} = \hat{F}_a(\eta, \xi) \tag{5.9}$$

where

$$\nabla^4_{\eta\xi} = \frac{\partial^4}{\partial \eta^4} + 2\frac{\partial^4}{\partial \eta^2 \partial \xi^2} + \frac{\partial^4}{\partial \xi^2} \tag{5.10}$$

and

$$\begin{aligned} K_F &= \frac{k_f a_o^4}{D} \qquad & \hat{N}_\eta &= \frac{N_x a_o^2}{D} \\ \hat{N}_\xi &= \frac{N_y a_o^2}{D} \qquad & \hat{F}_a &= \frac{F_a}{F_{max}} \end{aligned} \tag{5.11}$$

are nondimensional quantities.

The shear forces given by Eq. (5.7) become

$$\begin{aligned} \hat{V}_\eta(\eta, \xi) &= \frac{V_x}{F_{max}a_o} = \frac{\partial^3 \hat{w}}{\partial \eta^3} + (2 - \nu)\frac{\partial^3 \hat{w}}{\partial \eta \partial \xi^2} \\ \hat{V}_\xi(\eta, \xi) &= \frac{V_y}{F_{max}a_o} = \frac{\partial^3 \hat{w}}{\partial \xi^3} + (2 - \nu)\frac{\partial^3 \hat{w}}{\partial \xi \partial \eta^2} \end{aligned} \tag{5.12}$$

and the moments given by Eq. (5.6) become

$$\begin{aligned} \hat{M}_\eta(\eta, \xi) &= \frac{M_x}{F_{max}a_o^2} = -\left(\frac{\partial^2 \hat{w}}{\partial \eta^2} + \nu\frac{\partial^2 \hat{w}}{\partial \xi^2}\right) \\ \hat{M}_\xi(\eta, \xi) &= \frac{M_y}{F_{max}a_o^2} = -\left(\frac{\partial^2 \hat{w}}{\partial \xi^2} + \nu\frac{\partial^2 \hat{w}}{\partial \eta^2}\right) \end{aligned} \tag{5.13}$$

With these definitions, the boundary conditions given in Table 5.1 become those shown in Table 5.2, where in the table $\beta_o = b_o/a_o$. The applied loading given by Eq. (5.4) becomes

$$\hat{F}_a(\eta, \xi) = \hat{f}_\eta(\eta)\hat{f}_\xi(\xi) \tag{5.14}$$

Table 5.2 Several common boundary conditions for a rectangular plate: nondimensional form

Boundary	$\eta = 0$ or 1	$\xi = 0$ or β_o
Clamped	$w(\eta,\xi)\big\vert_{\eta=0,1} = 0$	$w(\eta,\xi)\big\vert_{\xi=0,\beta_o} = 0$
	$\frac{\partial w(\eta,\xi)}{\partial \eta}\Big\vert_{\eta=0,1} = 0$	$\frac{\partial w(\eta,\xi)}{\partial \xi}\Big\vert_{\xi=0,\beta_o} = 0$
Hinged	$w(\eta,\xi)\big\vert_{\eta=0,1} = 0$	$w(\eta,\xi)\big\vert_{\xi=0,\beta_o} = 0$
	$M_\eta(\eta,\xi)\big\vert_{\eta=0,1} = 0$	$M_\xi(\eta,\xi)\big\vert_{\xi=0,\beta_o} = 0$
Free	$M_\eta(\eta,\xi)\big\vert_{\eta=0,1} = 0$	$M_\xi(\eta,\xi)\big\vert_{\xi=0,\beta_o} = 0$
	$V_\eta(\eta,\xi)\big\vert_{\eta=0,1} + N_\eta \frac{\partial w(\eta,\xi)}{\partial \eta}\Big\vert_{\eta=0,1} = 0$	$V_\xi(\eta,\xi)\big\vert_{\xi=0,\beta_o} + N_\xi \frac{\partial w(\eta,\xi)}{\partial \xi}\Big\vert_{\xi=0,\beta_o} = 0$

where

$$
\begin{aligned}
\hat{f}_\xi(\xi) &= u(\xi - \xi_L) - u(\xi - \xi_R) \\
\hat{f}_\eta(\eta) &= (\hat{a}_p\eta + \hat{b}_p)\big(u(\eta - \eta_L) - u(\eta - \eta_R)\big)
\end{aligned}
\tag{5.15}
$$

and

$$
\begin{aligned}
\hat{a}_p &= \frac{q_L - q_R}{\eta_L - \eta_R} \\
\hat{b}_p &= \frac{q_R\eta_L - q_L\eta_R}{\eta_L - \eta_R}
\end{aligned}
\tag{5.16}
$$

In Eqs. (5.15) and (5.16), $\xi_L = b_L/a_o$, $\xi_R = b_R/a_o$, $\eta_L = a_L/a_o$, $\eta_R = a_R/a_o$, $q_L = F_L/F_{max}$, and $q_R = F_R/F_{max}$.

5.2 Solution for a Plate with Two Opposite Edges Hinged

Equation (5.9) is, in general, not separable; that is, one cannot find a solution of the form $\hat{w}(\eta,\xi) = X(\eta)Y(\xi)$. However, it has be shown[2] that if two opposite edges are hinged, then a separable solution exists. We shall assume that the edges $\xi = 0$ and $\xi = \beta_o$ are hinged and that there exists a solution of the form

$$
\hat{w}(\eta,\xi) = \sum_{m=1}^{\infty} X_m(\eta)\sin(\gamma_m\xi) \tag{5.17}
$$

[2] See [2] in Chapter 2, pp. 368–371.

where $\gamma_m = m\pi/\beta_o$. It is seen that at $\xi = 0$ and $\xi = \beta_o$, $\sin(\gamma_m\xi) = 0$. We note that $\sin(\gamma_m\xi)$ is an orthogonal function on the interval $0 \le \xi \le \beta_o$; that is,

$$\int_0^{\beta_o} \sin(\gamma_n\xi)\sin(\gamma_m\xi)d\xi = \frac{\beta_o}{2}\delta_{nm}, \tag{5.18}$$

where δ_{nm} is the Kronecker delta.

From Table 5.2, the hinged boundary conditions at the edges $\xi = 0$ and $\xi = \beta_o$ are

$$\begin{aligned} \hat{w}(\eta,\xi)\big|_{\xi=0,\beta_o} &= 0 \\ \hat{M}_\xi(\eta,\xi)\Big|_{\xi=0,\beta_o} &= -\left(\frac{\partial^2\hat{w}}{\partial\xi^2} + \nu\frac{\partial^2\hat{w}}{\partial\eta^2}\right)_{\xi=0,\beta_o} = 0 \end{aligned} \tag{5.19}$$

Upon substituting Eq. (5.17) in the hinged boundary condition given by Eq. (5.19), we see by inspection that the displacement boundary condition is satisfied and the moment boundary condition becomes

$$\hat{M}_\xi(\eta,\xi) = \sum_{m=1}^{\infty}\left(\gamma_m^2 X_m(\eta) - \nu\frac{\partial^2 X_m(\eta)}{\partial\eta^2}\right)\sin(\gamma_m\xi) \tag{5.20}$$

which equals zero when $\xi = 0$ and $\xi = \beta_o$.

The remaining boundary conditions on the edges $\eta = 0$ and $\eta = 1$ are obtained by substituting Eq. (5.17) into the column labeled "$\eta = 0$ or 1" of Table 5.2. This results in the following boundary conditions, where the prime denotes the derivative with respect to η.

Clamped edge: $\eta = 0$ and/or $\eta = 1$

$$\begin{aligned} X_m(\eta) &= 0 \\ X_m'(\eta) &= 0 \end{aligned} \tag{5.21}$$

Hinged edge: $\eta = 0$ or $\eta = 1$

$$\begin{aligned} X_m(\eta) &= 0 \\ -X_m''(\eta) + \nu\gamma_m^2 X_m(\eta) &= 0 \quad \rightarrow \quad X_m''(\eta) = 0 \end{aligned} \tag{5.22}$$

where we have used the first equation to simplify the second equation. Notice that this simplification mimics the moment boundary condition for a hinged end of a beam.

Free edge: $\eta = 0$ **and/or** $\eta = 1$

$$\begin{aligned} -X_m''(\eta) + \nu\gamma_m^2 X_m(\eta) &= 0 \\ X_m'''(\eta) + \left(\hat{N}_\eta - \gamma_m^2(2-\nu)\right)X_m'(\eta) &= 0 \end{aligned} \tag{5.23}$$

Returning to Eq. (5.9), Eq. (5.17) is substituted into Eq. (5.9) to obtain

$$\sum_{m=1}^{\infty} T_m(\eta)\sin(\gamma_m\xi) = \hat{f}_\eta(\eta)\hat{f}_\xi(\xi) \tag{5.24}$$

where

$$T_m(\eta) = \frac{d^4X_m}{d\eta^4} + u_{o,m}\frac{d^2X_m}{d\eta^2} + u_{f,m}X_m \tag{5.25}$$

and

$$\begin{aligned} u_{o,m} &= \hat{N}_\eta - 2\gamma_m^2 \\ u_{f,m} &= \gamma_m^4 - N_\xi\gamma_m^2 + K_F \end{aligned} \tag{5.26}$$

Multiplying Eq. (5.24) by the orthogonal function $\sin(\gamma_n\xi)$, integrating from $\xi = 0$ to $\xi = \beta_o$, and using Eq. (5.18), we arrive at

$$\frac{d^4X_m}{d\eta^4} + u_{o,m}\frac{d^2X_m}{d\eta^2} + u_{f,m}X_m = h_m\hat{f}_\eta(\eta) \tag{5.27}$$

where

$$\begin{aligned} h_m &= \frac{2}{\beta_o}\int_0^{\beta_o} \hat{f}_\xi(\xi)\sin(\gamma_m\xi)d\xi \\ &= \frac{2}{m\pi}\left(\cos(\gamma_m\xi_L) - \cos(\gamma_m\xi_R)\right) \end{aligned} \tag{5.28}$$

It is seen that Eq. (5.27) has the same form as Eq. (2.5), where $u_{o,m}$ corresponds to S_o, $u_{f,m}$ corresponds to K_f, $g_1 = \hat{f}_\eta(\eta)$, and when in Eq. (2.5) $F_{c,j} = T_j = 0$, $j = 1, 2, 3$, and $g_j = 0$, $j = 2, 3$. Then, from Eqs. (2.12)–(2.14), the solution to Eq. (5.27) is

$$\begin{aligned} X_m(\eta) = d'_{p,m}(\eta)X_m(0) &+ d_{p,m}(\eta)X_m'(0) + v'_{p,m}(\eta, 0)X_m''(0) \\ &+ v_{p,m}(\eta, 0)X_m'''(0) + h_m H_{p,m}(\eta) \end{aligned} \tag{5.29}$$

where

$$H_{p,m}(\eta) = \hat{a}_p A_{p,m}(\eta) + \hat{b}_p B_{p,m}(\eta) \tag{5.30}$$

and

$$\begin{aligned} A_{p,m}(\eta) &= \left(\eta_L r'_{p,m}(\eta, \eta_L) + r_{p,m}(\eta, \eta_L)\right)\theta(\eta - \eta_L) \\ &\quad - \left(\eta_R r'_{p,m}(\eta, \eta_R) + r_{p,m}(\eta, \eta_R)\right)\theta(\eta - \eta_R) \\ B_{p,m}(\eta) &= r'_{p,m}(\eta, \eta_L)\theta(\eta - \eta_L) - r'_{p,m}(\eta, \eta_R)\theta(\eta - \eta_R) \end{aligned} \tag{5.31}$$

In Eqs. (5.29)–(5.31), $r_{p,m}$, $v_{p,m}$, and $d_{p,m}$, respectively, are given in Tables 2.2, 2.3, and 2.4 with S_o replaced by $u_{o,m}$ and K_f replaced by $u_{f,m}$. If the edge at $\eta = 0$ is either clamped or hinged, then two of the solution constants are zero. The remaining two solution constants are determined from the boundary conditions at $\eta = 1$. The procedure for obtaining these constants is virtually the same as that employed in Chap. 2.

5.3 Numerical Results: A Special Case

To illustrate the methodology, we consider the case where $\hat{N}_\eta = \hat{N}_\xi = K_F = 0$. Then $S_o = u_{o,m} = -2\gamma_m^2$ and $K_f = u_{f,m} = \gamma_m^4$. From the definitions in Tables 2.2, 2.3, and 2.4, we find that $a = \gamma_m^2$ and $b = 0$. With these definitions and by taking the limits as $b \to 0$ in the expressions in the column of these tables for which $S_o \neq 0$ and $K_f > 0$, we obtain the following:

$$\begin{aligned} r_{p,m}(\eta, \alpha) &= \frac{1}{\gamma_m^4}(\eta - \alpha)\left(1 + \frac{1}{2}\cosh\left(\gamma_m(\eta - \alpha)\right)\right) \\ &\quad - \frac{3}{2\gamma_m^5}\sinh\left(\gamma_m(\eta - \alpha)\right) \\ v_{p,m}(\eta, \alpha) &= \frac{1}{2\gamma_m^2}(\eta - \alpha)\cosh\left(\gamma_m(\eta - \alpha)\right) - \frac{1}{2\gamma_m^3}\sinh\left(\gamma_m(\eta - \alpha)\right) \\ d_{p,m}(\eta) &= \frac{1}{2\gamma_m}\left(-\eta\gamma_m\cosh(\gamma_m\eta) + \sinh(\gamma_m\eta)\right) \end{aligned} \tag{5.32}$$

The derivatives of these functions are given in Table 5.3.

For the boundary conditions, we assume that the edge $\eta = 0$ is clamped and that the edge $\eta = 1$ is free. Then, Eq. (5.29) becomes

$$X_m(\eta) = v'_{p,m}(\eta, 0)X''_m(0) + v_{p,m}(\eta, 0)X'''_m(0) + h_m H_{p,m}(\eta) \tag{5.33}$$

Table 5.3 Derivatives of the functions in Eq. (5.32)

$\Psi(\eta) =$	$r_{p,m}(\eta,\alpha)$	$v_{p,m}(\eta,0)$	$d_{p,m}(\eta)$
$\Psi'(\eta)$	$\frac{1}{\gamma_m^4}\big(1-\cosh(\gamma_m(\eta-\alpha))\big)$ $+\frac{(\eta-\alpha)}{\gamma_m^3}\sinh\big(\gamma_m(\eta-\alpha)\big)$	$\frac{\eta}{2\gamma_m}\sinh(\gamma_m\eta)$	$-\frac{\gamma_m\eta}{2}\sinh(\gamma_m\eta)$
$\Psi''(\eta)$	$\frac{\eta-\alpha}{2\gamma_m^2}\cosh\big(\gamma_m(\eta-\alpha)\big)$ $-\frac{1}{2\gamma_m^3}\sinh\big(\gamma_m(\eta-\alpha)\big)$	$\frac{\eta}{2}\cosh(\gamma_m\eta)$ $+\frac{1}{2\gamma_m}\sinh(\gamma_m\eta)$	$-\frac{\gamma_m^2\eta}{2}\cosh(\gamma_m\eta)$ $-\frac{\gamma_m}{2}\sinh(\gamma_m\eta)$
$\Psi'''(\eta)$	$\frac{\eta-\alpha}{2\gamma_m}\sinh\big(\gamma_m(\eta-\alpha)\big)$	$\cosh(\gamma_m\eta)$ $+\frac{\gamma_m\eta}{2}\sinh(\gamma_m\eta)$	$-\gamma_m^2\cosh(\gamma_m\eta)$ $-\frac{\gamma_m^3\eta}{2}\sinh(\gamma_m\eta)$
$\Psi''''(\eta)$	$\frac{\eta-\alpha}{2}\cosh\big(\gamma_m(\eta-\alpha)\big)$ $+\frac{1}{2\gamma_m}\sinh\big(\gamma_m(\eta-\alpha)\big)$	$\frac{\gamma_m^2\eta}{2}\cosh(\gamma_m\eta)$ $+\frac{3\gamma_m}{2}\sinh(\gamma_m\eta)$	$-\frac{\gamma_m^4\eta}{2}\cosh(\gamma_m\eta)$ $-\frac{3\gamma_m^3}{2}\sinh(\gamma_m\eta)$

Substituting Eq. (5.33) into Eq. (5.23) with $\hat{N}_\eta = 0$ and setting $\eta = 1$, we obtain

$$\begin{aligned} s'_{1,m}(1)X''_m(0) + s_{1,m}(1)X'''_m(0) &= G_{1,m}(1) \\ s'_{2,m}(1)X''_m(0) + s_{2,m}(1)X'''_m(0) &= G_{2,m}(1) \end{aligned} \tag{5.34}$$

where

$$\begin{aligned} s_{1,m}(\eta) &= \nu\gamma_m^2 v_{p,m}(\eta,0) - v''_{p,m}(\eta,0) \\ s_{2,m}(\eta) &= -\gamma_m^2(2-\nu)v'_{p,m}(\eta,0) + v'''_{p,m}(\eta,0) \end{aligned} \tag{5.35}$$

and

$$\begin{aligned} G_{1,m}(\eta) =& h_m\Big(-\nu\gamma_m^2 H_{p,m}(\eta) + H''_{p,m}(\eta)\Big) \\ G_{2,m}(\eta) =& h_m\Big(\gamma_m^2(2-\nu)H'_{p,m}(\eta) - H'''_{p,m}(\eta)\Big) \end{aligned} \tag{5.36}$$

Solving Eq. (5.34) for $X''_m(0)$ and $X'''_m(0)$ yields

$$\begin{aligned} X''_m(0) =& \frac{1}{D_{p,m}}\Big(s_{2,m}(1)G_{1,m}(1) - s_{1,m}(1)G_{2,m}(1)\Big) \\ X'''_m(0) =& \frac{1}{D_{p,m}}\Big(s'_{1,m}(1)G_{2,m}(1) - s'_{2,m}(1)G_{1,m}(1)\Big) \end{aligned} \tag{5.37}$$

where

$$D_{p,m} = s'_{1,m}(1)s_{2,m}(1) - s'_{2,m}(1)s_{1,m}(1) \tag{5.38}$$

Substituting Eq. (5.37) into Eq. (5.33) and this result, in turn, into Eq. (5.17) gives the displacement for the surface of the plate. We then use this expression for the displacement in Eq. (5.13) to obtain Fig. 5.2, which was obtained for the values shown in the figure legend. Six terms in the series were used to obtain these results.

In Fig. 5.2, the image on the top left shows the location and shape of the distributed load. The image on the top right shows a contour plot of the displacement $\hat{w}(\eta, \xi)$ and its maximum value and the location where it occurs. The two bottom images show contour plots of the moments $\hat{M}_\eta(\eta, \xi)$ and $\hat{M}_\xi(\eta, \xi)$ and their respective maximum and minimum values and the locations where they occur.

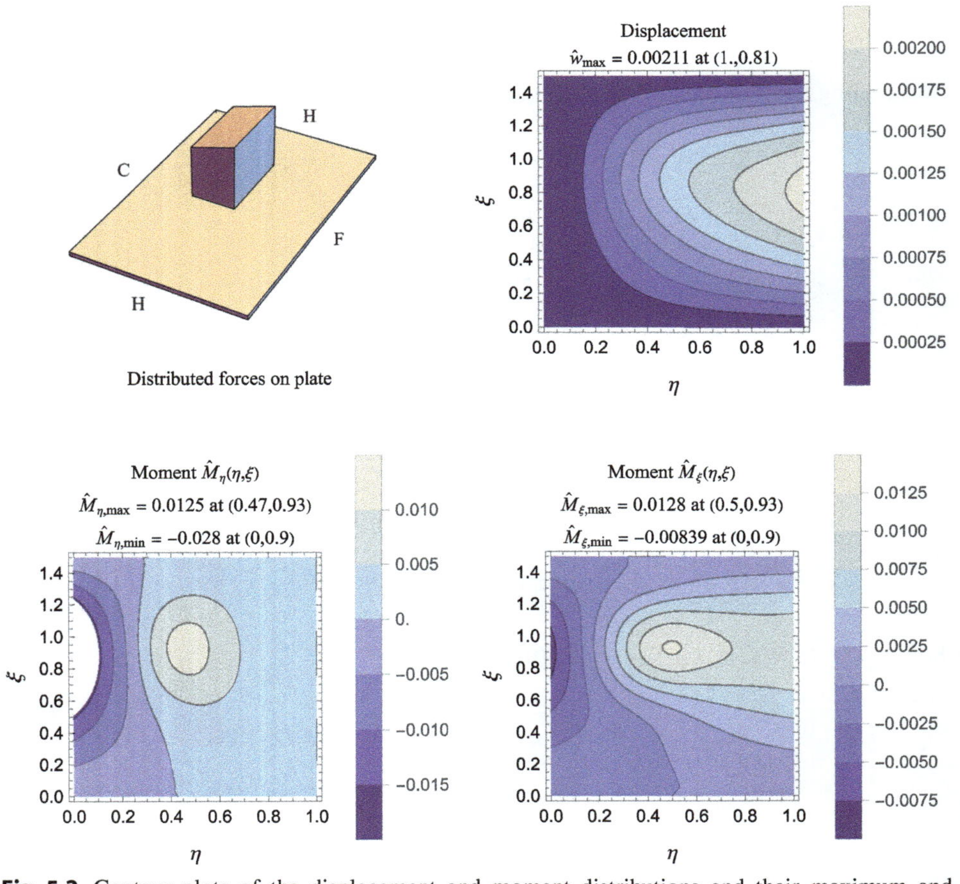

Fig. 5.2 Contour plots of the displacement and moment distributions and their maximum and minimum values in a rectangular plate when $\beta_o = 1.5$, $q_L = 0.8$, $q_R = 1.0$, $\eta_L = 0.3$, $\eta_R = 0.55$, $\xi_L = 0.7$, $\xi_R = 1.15$, and $\nu = 0.3$

Circular Arches

6

Abstract

In this chapter, we show how the solution method employed in Chap. 2 can be used to obtain the static response of an asymmetrical circular arch that is subject to vertical, radial, and gravity forces.

Keywords

Circular arch · Radial forces · Tangential forces · Gravity forces · Boundary conditions

6.1 Governing Equations

Consider the uniform circular arch shown in Fig. 6.1, which has a radius R_o (m), a cross-sectional area A (m^2), a moment of inertia I (m^4), and a tensile modulus E (N/m^2). The line representing R_o is vertical. Then the origin of the arch, that is, $\varphi = 0$, is a counterclockwise angle α from the line denoted R_o. The arch spans an angle β_a, where $\alpha < \beta_a \leq 2\alpha$. If $\bar{w} = \bar{w}(\varphi)$ is the radial displacement and $\bar{v} = \bar{v}(\varphi)$ is the tangential displacement of the arch, then the differential equations governing for this system are [1,2].

Supplementary Information The online version contains supplementary material available at (https://doi.org/10.1007/978-3-032-28726-7_6).

E. B. Magrab, *Static Analysis of Beams, Frames, and Arches*, Synthesis Lectures on Mechanical Engineering, https://doi.org/10.1007/978-3-032-28726-7_6

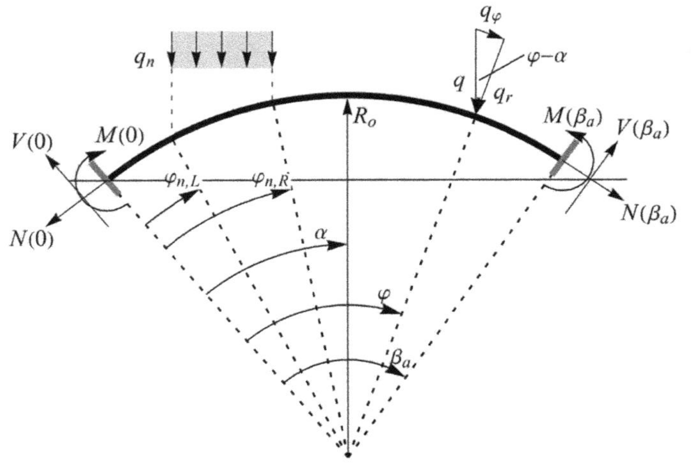

Fig. 6.1 Nomenclature for a circular arch of radius R_o with both ends fixed

$$\begin{aligned}\frac{EA}{R_o^2}\left(\frac{d^2\bar{v}}{d\varphi^2}-\frac{d\bar{w}}{d\varphi}\right)+\frac{EI}{R_o^4}\left(\frac{d^3\bar{w}}{d\varphi^3}+\frac{d^2\bar{v}}{d\varphi^2}\right)+\bar{Q}_\varphi(\varphi)=0\\ \frac{EA}{R_o^2}\left(\frac{d\bar{v}}{d\varphi}-\bar{w}\right)-\frac{EI}{R_o^4}\left(\frac{d^4\bar{w}}{d\varphi^4}+\frac{d^3\bar{v}}{d\varphi^3}\right)+\bar{Q}_r(\varphi)=0\end{aligned} \tag{6.1}$$

where $\bar{Q}_\varphi(\varphi)$ (N/m) is the tangential component and $\bar{Q}_r(\varphi)$ (N/m) is the radial component of the applied forces.

In terms of displacements, the bending moment is given by

$$\bar{M}(\varphi)=-\frac{EI}{R_o^2}\left(\frac{d^2\bar{w}}{d\varphi^2}+\frac{d\bar{v}}{d\varphi}\right)\quad \text{Nm} \tag{6.2}$$

the axial force by

$$\bar{N}(\varphi)=-\frac{EA}{R_o}\left(\frac{d\bar{v}}{d\varphi}-\bar{w}\right)\quad \text{N} \tag{6.3}$$

and the radial transverse shear force by

$$\bar{V}(\varphi)=\frac{1}{R_o}\frac{d\bar{M}(\varphi)}{d\varphi}=-\frac{EI}{R_o^3}\left(\frac{d^3\bar{w}}{d\varphi^3}+\frac{d^2\bar{v}}{d\varphi^2}\right)\quad \text{N} \tag{6.4}$$

These quantities are depicted at the ends of the arch in Fig. 6.1.

The boundary conditions that we shall consider at each end of the arch are as follows.

Clamped at $\varphi = 0$ and/or $\varphi = \beta_a$

$$\bar{w}(\varphi) = \bar{v}(\varphi) = \bar{w}'(\varphi) = 0 \tag{6.5}$$

Hinged at $\varphi = 0$ and/or $\varphi = \beta_a$

$$\bar{w}(\varphi) = \bar{v}(\varphi) = \bar{M}(\varphi) = 0 \tag{6.6}$$

In Eq. (6.5), the prime denotes the derivative with respect to φ.

There are three types of applied forces that we shall consider. Referring to Fig. 6.1, the first is a set of nonoverlapping spatially distributed vertical forces of the form

$$\begin{aligned}\bar{Q}_\varphi(\varphi) &\to Q_{\varphi,y}(\varphi) = \sum_{n=1}^{N_y} q_n \sin(\varphi - \alpha)\big(u(\varphi - \varphi_{n,L}) - u(\varphi - \varphi_{n,R})\big) \\ \bar{Q}_r(\varphi) &\to Q_{r,y}(\varphi) = \sum_{n=1}^{N_y} q_n \cos(\varphi - \alpha)\big(u(\varphi - \varphi_{n,L}) - u(\varphi - \varphi_{n,R})\big)\end{aligned} \tag{6.7}$$

where $u(\varphi)$ is the unit step function, q_n (N/m) is the magnitude of the vertical force of the nth load segment, N_y is the number of vertical load segments, and $\varphi_{n,L}$ and $\varphi_{n,R}$ denote the angular limits of the nth load segment. Refer to Fig. 6.1.

The second type of applied force is one that only acts radially. In this case, Eq. (6.7) becomes

$$\begin{aligned}\bar{Q}_\varphi(\varphi) &\to Q_{\varphi,r}(\varphi) = 0 \\ \bar{Q}_r(\varphi) &\to Q_{r,r}(\varphi) = \sum_{n=1}^{N_r} q_{r,n}\big(u(\varphi - \varphi_{r,n,L}) - u(\varphi - \varphi_{r,n,R})\big)\end{aligned} \tag{6.8}$$

where $q_{r,n}$ (N/m) is the magnitude of the radial force of the nth segment acting over the range $\varphi_{r,n,L}$ to $\varphi_{r,n,R}$ and N_r is the number of load segments.

The third case is the force due to gravity. If the density of the uniform arch is ρ (kg/m^3) and $g = 9.8$ m/s^2 is the gravitational constant, then the force per unit length of the arch due to gravity is $q_g = \rho A g$ (N/m), and the tangential and radial components, respectively, of this force are

$$\begin{aligned}\bar{Q}_\varphi(\varphi) &\to Q_{\varphi,g}(\varphi) = q_g \sin(\varphi - \alpha) \\ \bar{Q}_r(\varphi) &\to Q_{r,g}(\varphi) = q_g \cos(\varphi - \alpha)\end{aligned} \tag{6.9}$$

It is noted that when $N_y = 1$, $\varphi_{1,L} = 0$, and $\varphi_{1,L} = \beta_a$ in Eq. (6.7), Eq. (6.7) is the same as Eq. (6.9) when $q_g = q_1$.

To place the governing equations and boundary conditions into a nondimensional form, we introduce the parameters

$$w = \frac{EA\bar{w}}{F_{max}R_o^2} \qquad v = \frac{EA\bar{v}}{F_{max}R_o^2} \tag{6.10}$$

where F_{max} (N/m) is the maximum of q_n, $q_{r,n}$, and q_g. Then, Eq. (6.1) becomes

$$\begin{aligned} &(1+\lambda)\frac{d^2v}{d\varphi^2} + \lambda\frac{d^3v}{d\varphi^3} - \frac{dw}{d\varphi} + \hat{Q}_\varphi(\varphi) = 0 \\ &\frac{dv}{d\varphi} - w - \lambda\left(\frac{d^4w}{d\varphi^4} + \frac{d^3v}{d\varphi^3}\right) + \hat{Q}_r(\varphi) = 0 \end{aligned} \tag{6.11}$$

In Eq. (6.11), $\lambda = r^2/R_o^2$, $r = \sqrt{I/A}$ is the radius of gyration of the cross section, and

$$\begin{aligned} \hat{Q}_\varphi(\varphi) &= \hat{Q}_{\varphi,y}(\varphi) + \hat{Q}_{\varphi,g}(\varphi) \\ \hat{Q}_r(\varphi) &= \hat{Q}_{r,y}(\varphi) + \hat{Q}_{r,r}(\varphi) + \hat{Q}_{r,g}(\varphi) \end{aligned} \tag{6.12}$$

where

$$\begin{aligned} \hat{Q}_{\varphi,y}(\varphi) &= \sum_{n=1}^{N_y} \hat{q}_n \sin(\varphi-\alpha)\big(u(\varphi-\varphi_{n,L}) - u(\varphi-\varphi_{n,R})\big) \\ \hat{Q}_{r,y}(\varphi) &= \sum_{n=1}^{N_y} \hat{q}_n \cos(\varphi-\alpha)\big(u(\varphi-\varphi_{n,L}) - u(\varphi-\varphi_{n,R})\big) \\ \hat{Q}_{r,r}(\varphi) &= \sum_{n=1}^{N_r} \hat{q}_{r,n}\big(u(\varphi-\varphi_{r,n,L}) - u(\varphi-\varphi_{r,n,R})\big) \\ \hat{Q}_{\varphi,g}(\varphi) &= \hat{q}_g \sin(\varphi-\alpha) \\ \hat{Q}_{r,g}(\varphi) &= \hat{q}_g \cos(\varphi-\alpha) \end{aligned} \tag{6.13}$$

and $\hat{q}_n = q_n/F_{max}$, $\hat{q}_{r,n} = q_{r,n}/F_{max}$, and $\hat{q}_g = q_g/F_{max}$.

The nondimensional forms for the bending moment, axial force, and the radial transverse shear force, respectively, are

$$
\begin{aligned}
M(\varphi) &= \frac{\bar{M}(\varphi)}{R_o^2 F_{max}} = -\lambda\left(\frac{d^2w}{d\varphi^2} + \frac{dv}{d\varphi}\right) \\
N(\varphi) &= \frac{\bar{N}(\varphi)}{R_o F_{max}} = -\left(\frac{dv}{d\varphi} - w\right) \\
V(\varphi) &= \frac{\bar{V}(\varphi)}{R_o F_{max}} = -\lambda\left(\frac{d^3w}{d\varphi^3} + \frac{d^2v}{d\varphi^2}\right)
\end{aligned}
\tag{6.14}
$$

The boundary conditions become as follows.

Clamped at $\varphi = 0$ and/or $\varphi = \beta_a$

$$
w(\varphi) = v(\varphi) = w'(\varphi) = 0 \tag{6.15}
$$

Hinged at $\varphi = 0$ and/or $\varphi = \beta_a$

$$
w(\varphi) = v(\varphi) = M(\varphi) = 0 \tag{6.16}
$$

In Eq. (6.15), the prime denotes the derivative with respect to φ.

The solution to Eq. (6.11) with the boundary conditions given by Eqs. (6.15) and (6.16) is obtained using the Laplace transform. If s is the Laplace transform variable and $\tilde{w}(s)$ is the Laplace transform of $w(\varphi)$ and $\tilde{v}(s)$ is the Laplace transform of $v(\varphi)$, then taking the Laplace transform of Eq. (6.11) with respect to φ, solving for $\tilde{w}(s)$ and $\tilde{v}(s)$, and then taking the inverse Laplace transform yield[1]

$$
\begin{aligned}
w(\varphi) = g_1(\varphi)V_0 &+ g_2(\varphi)v'(0) \\
&+ g_4'(\varphi)w''(0) + g_4''(\varphi)w'(0) \\
&+ \big(1 - g_2(\varphi)\big)w(0) + \hat{H}_r(\varphi)
\end{aligned}
\tag{6.17}
$$

and

$$
\begin{aligned}
v(\varphi) = \big(-g_1'(\varphi) &+ g_4'(\varphi)\big)V_0 + g_3(\varphi)v'(0) \\
&+ g_4(\varphi)w''(0) + g_4'(\varphi)w'(0) \\
&+ \big(\varphi - g_3(\varphi)\big)w(0) + v(0) + \hat{H}_\varphi(\varphi)
\end{aligned}
\tag{6.18}
$$

[1] These operations were performed using Mathematica®.

where $V_0 = w'''(0) + v''(0) = -V(0)$, the prime denotes the derivative with respect to φ,

$$\begin{aligned}
g_1(\varphi) &= \frac{1}{2}(1+\lambda)\big(\sin\varphi - \varphi\cos\varphi\big) \\
g_2(\varphi) &= \frac{1+\lambda}{\lambda}\left(1 - \frac{1}{2}\varphi\sin\varphi - \cos\varphi\right) \\
g_3(\varphi) &= \frac{1+\lambda}{\lambda}\varphi - \frac{3+\lambda}{2\lambda}\sin\varphi + \frac{1+\lambda}{2\lambda}\varphi\cos\varphi) \\
g_4(\varphi) &= \varphi - \sin\varphi
\end{aligned} \tag{6.19}$$

and

$$\begin{aligned}
\hat{H}_r(\varphi) &= \sum_{n=1}^{N_y} \hat{q}_n\Big(c_w(\varphi,\varphi_{n,L})u(\varphi-\varphi_{n,L}) - c_w(\varphi,\varphi_{n,R})u(\varphi-\varphi_{n,R})\Big) \\
&\quad + \sum_{n=1}^{N_r} \hat{q}_{r,n}\Big(g_2(\varphi-\varphi_{r,n,L})u(\varphi-\varphi_{r,n,L}) \\
&\qquad - g_2(\varphi-\varphi_{r,n,R})u(\varphi-\varphi_{r,n,R})\Big) + \hat{q}_g c_w(\varphi,0) \\
\hat{H}_\varphi(\varphi) &= \sum_{n=1}^{N_y} \hat{q}_n\Big(c_v(\varphi,\varphi_{n,L})u(\varphi-\varphi_{n,L}) - c_v(\varphi,\varphi_{n,R})u(\varphi-\varphi_{n,R})\Big) \\
&\quad + \sum_{n=1}^{N_r} \hat{q}_{r,n}\Big(\big(g_3(\varphi-\varphi_{r,n,L}) - (\varphi-\varphi_{r,n,L})\big)u(\varphi-\varphi_{r,n,L}) \\
&\qquad - \big(g_3(\varphi-\varphi_{r,n,R}) - (\varphi-\varphi_{r,n,R})\big)u(\varphi-\varphi_{r,n,R})\Big) \\
&\qquad + \hat{q}_g c_v(\varphi,0)
\end{aligned} \tag{6.20}$$

In Eq. (6.20),

$$\begin{aligned}
c_w(\varphi,x) &= \frac{1}{8\lambda}\Big(\Big(7+\lambda-2(\varphi-x)^2(1+\lambda)\Big)\cos(\varphi-\alpha) \\
&\qquad - 8\cos(x-\alpha) + (1-\lambda)\cos(\varphi-2x+\alpha) \\
&\qquad + 2(\varphi-x)(3+\lambda)\sin(\varphi-\alpha)\Big) \\
c_v(\varphi,x) &= \frac{1}{8\lambda}\Big(2(\varphi-x)(\lambda-5)\cos(\varphi-\alpha) \\
&\qquad - 8(\varphi-x)\cos(x-\alpha) - 16\sin(x-\alpha) \\
&\qquad + \Big(17-\lambda-2(\varphi-x)^2(1+\lambda)\Big)\sin(\varphi-\alpha) \\
&\qquad + (1-\lambda)\sin(\varphi-2x+\alpha)\Big)
\end{aligned} \tag{6.21}$$

It is noted that $c_w(x, x) = 0$ as does its first two derivatives and $c_v(x, x) = 0$ as does its first derivative.

We now use the boundary conditions given by Eqs. (6.15) and (6.16) to simplify Eqs. (6.17) and (6.18). When the end at $\varphi = 0$ is clamped, we use Eq. (6.15) to obtain

$$\begin{aligned} w_c(\varphi) &= g_1(\varphi)V_0 + g_2(\varphi)v'(0) \\ &\quad + g_4'(\varphi)w''(0) + \hat{H}_r(\varphi) \end{aligned} \tag{6.22}$$

and

$$\begin{aligned} v_c(\varphi) &= \big(- g_1'(\varphi) + g_4'(\varphi)\big)v'(0) \\ &\quad + g_4(\varphi)w''(0) + \hat{H}_\varphi(\varphi) \end{aligned} \tag{6.23}$$

The three unknown solution constants V_0, $v'(0)$, and $w''(0)$ are determined from the boundary conditions at $\varphi = \beta_a$.

When the end at $\varphi = 0$ is hinged, we use Eq. (6.16) to obtain

$$\begin{aligned} w_h(\varphi) &= g_1(\varphi)V_0 + \big(g_2(\varphi) - g_4'(\varphi)\big)\, v'(0) \\ &\quad + g_4''(\varphi)w'(0) + \hat{H}_r(\varphi) \end{aligned} \tag{6.24}$$

and

$$\begin{aligned} v_h(\varphi) &= \big(- g_1'(\varphi) + g_4'(\varphi)\big)V_0 + \big(g_3(\varphi) - g_4(\varphi)\big)v'(0) \\ &\quad + g_4'(\varphi)w'(0) + \hat{H}_\varphi(\varphi) \end{aligned} \tag{6.25}$$

where we have used the fact that

$$M(0) = -w''(0) - v'(0) = 0 \rightarrow w''(0) = -v'(0) \tag{6.26}$$

The three unknown solution constants V_0, $v'(0)$, and $w'(0)$ are determined from the boundary conditions at $\varphi = \beta_a$.

Arch clamped at $\boldsymbol{\varphi = 0}$ and $\boldsymbol{\varphi = \beta_a}$

For an arch clamped at both ends, we use Eqs. (6.22) and (6.23) in Eq. (6.15) and obtain

$$\mathbf{A}_c\mathbf{c}_c = \mathbf{b}_c \tag{6.27}$$

where

$$\mathbf{A}_c = \begin{pmatrix} g_1(\beta) & g_2(\beta) & g_4'(\beta) \\ -g_1'(\beta) + g_4'(\beta) & g_3(\beta) & g_4(\beta) \\ g_1'(\beta) & g_2'(\beta) & g_4''(\beta) \end{pmatrix} \tag{6.28}$$

and

$$\mathbf{c}_c = \begin{Bmatrix} V_0 \\ v'(0) \\ w''(0) \end{Bmatrix} \qquad \mathbf{b}_c = \begin{Bmatrix} -\hat{H}_r(\beta) \\ -\hat{H}_\varphi(\beta) \\ -\hat{H}_r'(\beta) \end{Bmatrix} \tag{6.29}$$

Upon the numerical evaluation of Eq. (6.27), the values from $\mathbf{c}_c$ are substituted into Eqs. (6.22) and (6.23), which are then used in Eq. (6.14) to obtain the moment distribution, axial force, and shear force.

Arch hinged at $\varphi = 0$ and $\varphi = \beta$
For an arch hinged at both ends, we use Eqs. (6.24) and (6.25) in Eq. (6.16) and obtain

$$\mathbf{A}_h \mathbf{c}_h = \mathbf{b}_h \tag{6.30}$$

where

$$\mathbf{A}_h = \begin{pmatrix} g_1(\beta) & g_2(\beta) & g_4'(\beta) \\ -g_1'(\beta) + g_4'(\beta) & g_3(\beta) & g_4(\beta) \\ g_4''(\beta) & \begin{matrix} g_2''(\beta) + g_3'(\beta) \\ -g_4'(\beta) - g_4'''(\beta) \end{matrix} & g_4''(\beta) + g_4''''(\beta) \end{pmatrix} \tag{6.31}$$

and

$$\mathbf{c}_h = \begin{Bmatrix} V_0 \\ v'(0) \\ w'(0) \end{Bmatrix} \qquad \mathbf{b}_h = \begin{Bmatrix} -\hat{H}_r(\beta) \\ -\hat{H}_\varphi(\beta) \\ -\hat{H}_r''(\beta) - \hat{H}_\varphi'(\beta) \end{Bmatrix} \tag{6.32}$$

Upon the numerical evaluation of (6.30), the values from $\mathbf{c}_h$ are substituted into Eqs. (6.24) and (6.25), which are then used in Eq. (6.14) to obtain the moment distribution, the axial force, and the shear force.

6.2 Numerical Results: Arch Fixed at Both Ends

We illustrate the methodology by considering the case of an arch fixed at both ends. As indicated in Sect. 6.1, we use the values from $\mathbf{c}_c$ that have been obtained from Eq. (6.27). These values are then substituted into Eqs. (6.22) and (6.23), which, in turn, are then used in Eq. (6.14) to obtain the moment distribution, the axial force at the ends of the arch, and shear force at the ends of the arch.

In each of the figures, the image on the top left shows the radial displacement $w_c(\varphi)$ as a function of the angular position, the image on the top right shows the tangential

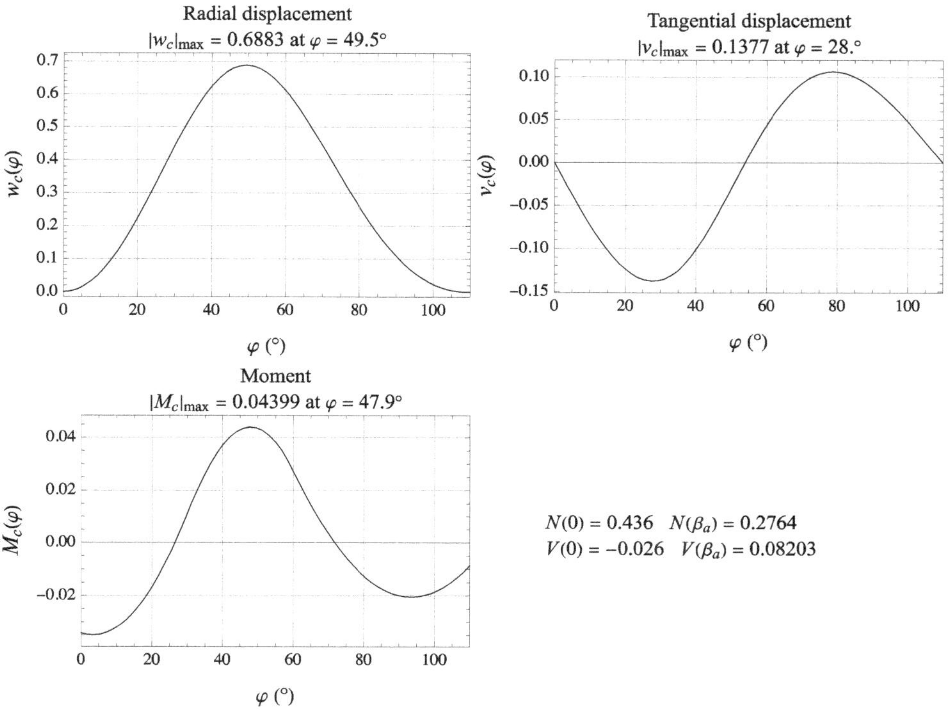

Fig. 6.2 Displacement and moment response of a circular arch fixed at both ends subjected to a vertical load applied over the range $\varphi_{1,L} = 30°$ to $\varphi_{1,R} = 60°$ for $\lambda = 0.01, \alpha = 60°$, and $\beta_a = 110°$

displacement $v_c(\varphi)$ as a function of the angular position, and the image on the bottom left shows the moment $M_c(\varphi)$ as a function of the angular position. Accompanying each of these images is their respective absolute maximum value and the angle at which it occurs. The values of the axial force and the radial transverse shear force at each end of the arch are shown in the image on the bottom right.

Three asymmetric loading cases are considered for an asymmetrical arch with $\alpha = 60°$, $\beta_a = 110°$, and $\hat{q}_1 = \hat{q}_{r,1} = \hat{q}_g = 1$. For the first case, a vertical load is applied over the range $\varphi_{1,L} = 30°$ to $\varphi_{1,R} = 60°$. The results are shown in Fig. 6.2 for $\lambda = 0.01$ and in Fig. 6.3 for $\lambda = 0.001$. For the second case, a radial load is applied in the range $\varphi_{r,1,L} = 30°$ to $\varphi_{r,1,L} = 60°$. The results are shown in Fig. 6.4 for $\lambda = 0.01$. For the last case, only the gravity load is considered. The results are shown in Fig. 6.5 for $\lambda = 0.01$.

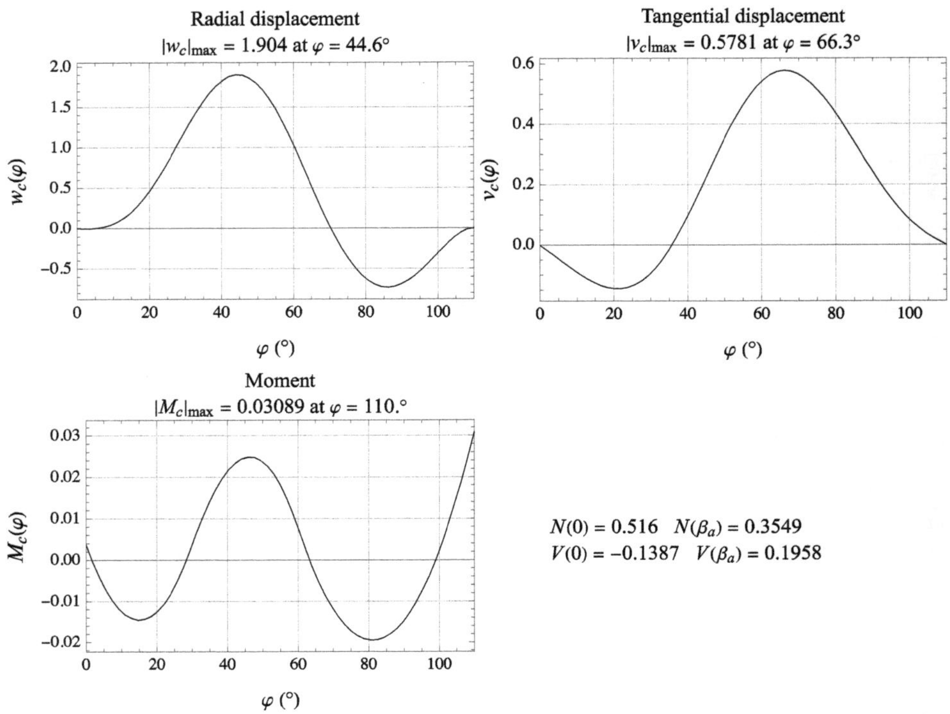

Fig. 6.3 Displacement and moment response of a circular arch fixed at both ends subjected to a vertical load applied over the range $\varphi_{1,L} = 30°$ to $\varphi_{1,R} = 60°$ for $\lambda = 0.001, \alpha = 60°$, and $\beta_a = 110°$

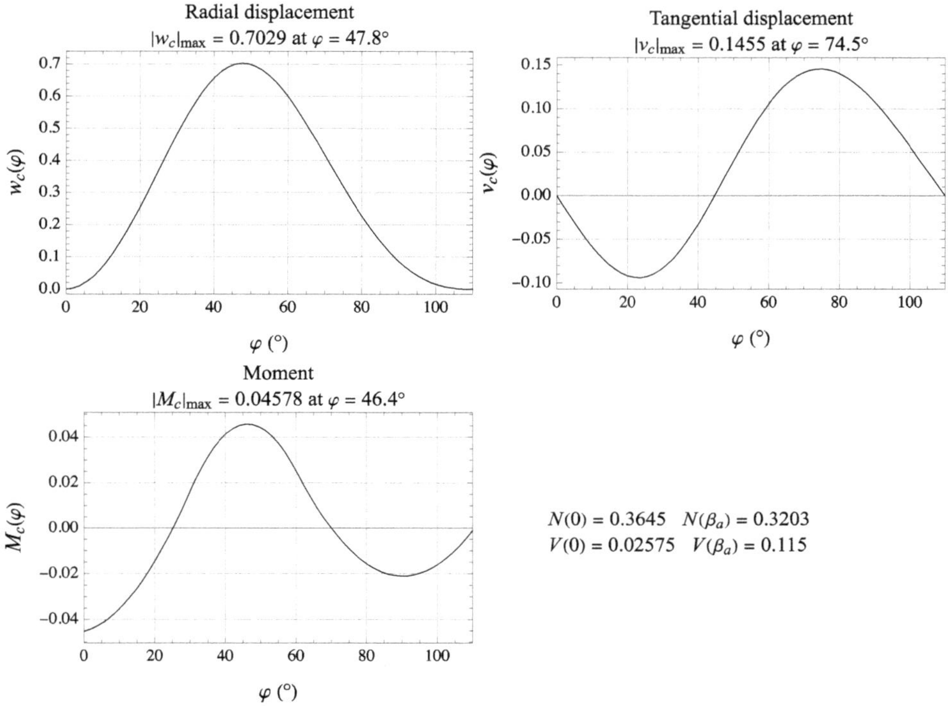

Fig. 6.4 Displacement and moment response of a circular arch fixed at both ends subjected to a radial load applied over the range $\varphi_{r,1,L} = 30°$ to $\varphi_{r,1,R} = 60°$ for $\lambda = 0.01, \alpha = 60°$, and $\beta_a = 110°$

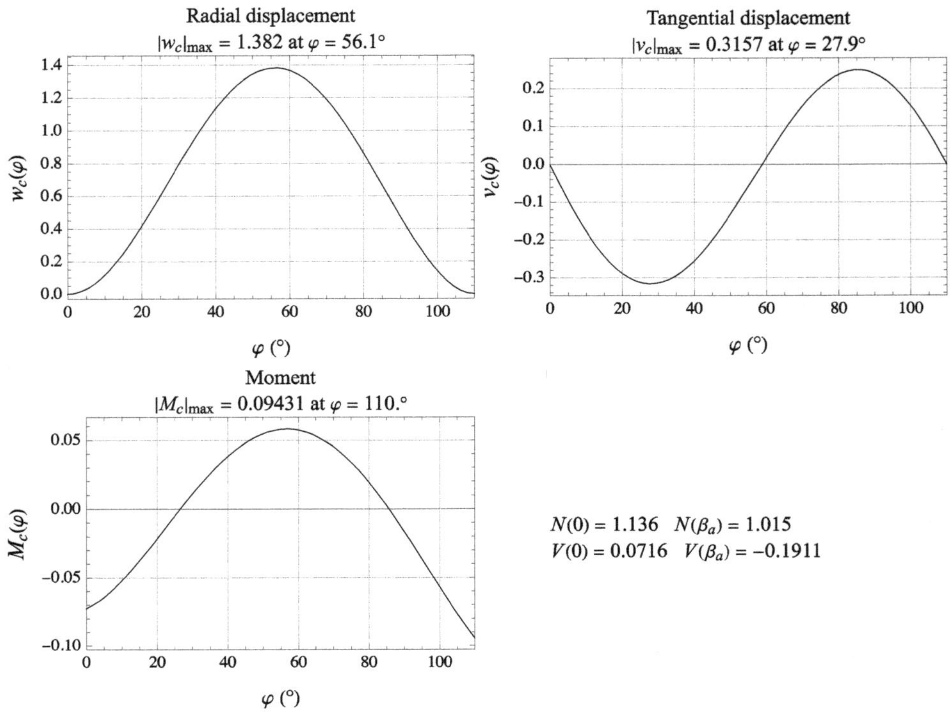

Fig. 6.5 Displacement and moment response of a circular arch fixed at both ends subjected to gravity for $\lambda = 0.01$, $\alpha = 60°$, and $\beta_a = 110°$

References

1. Dym CL, Williams HE (2011) Stress and displacement estimates for arches. ASCE J Struct Engrg 137(1):49–58
2. Glisic B (2020) New in old: simplified equations for linear-elastic symmetric arches and insights on their behavior. J Intl Assoc Shell Spatial Struct 61(3):227–240

GPSR Compliance
The European Union's (EU) General Product Safety Regulation (GPSR) is a set of rules that requires consumer products to be safe and our obligations to ensure this.

If you have any concerns about our products, you can contact us on

ProductSafety@springernature.com

In case Publisher is established outside the EU, the EU authorized representative is:

Springer Nature Customer Service Center GmbH
Europaplatz 3
69115 Heidelberg, Germany

www.ingramcontent.com/pod-product-compliance
Ingram Content Group UK Ltd.
Pitfield, Milton Keynes, MK11 3LW, UK
UKHW061927020926
475114UK00005B/71
* 9 7 8 3 0 3 2 2 8 7 2 5 0 *

GPSR Compliance
The European Union's (EU) General Product Safety Regulation (GPSR) is a set of rules that requires consumer products to be safe and our obligations to ensure this.

If you have any concerns about our products, you can contact us on

ProductSafety@springernature.com

In case Publisher is established outside the EU, the EU authorized representative is:

Springer Nature Customer Service Center GmbH
Europaplatz 3
69115 Heidelberg, Germany

www.ingramcontent.com/pod-product-compliance
Ingram Content Group UK Ltd.
Pitfield, Milton Keynes, MK11 3LW, UK
UKHW061927020926
475114UK00005B/71
* 9 7 8 3 0 3 2 2 8 7 2 5 0 *